AF779852

Influence of Urbanization-Related Radical Land Modification on Urban Extreme Climate

Influence of Urbanization-Related Radical Land Modification on Urban Extreme Climate

Guest Editors

Guoyu Ren
Yali Luo

MDPI

Basel • Beijing • Wuhan • Barcelona • Belgrade • Novi Sad • Cluj • Manchester

Guest Editors
Guoyu Ren
Department of Atmospheric Sciences
China University of Geosciences
Wuhan
China

Yali Luo
School of Atmospheric Sciences
Nanjing University of Information Science & Technology (NUIST)
Nanjing
China

Editorial Office
MDPI AG
Grosspeteranlage 5
4052 Basel, Switzerland

This is a reprint of the Special Issue, published open access by the journal *Land* (ISSN 2073-445X), freely accessible at: https://www.mdpi.com/journal/land/special_issues/urbanization_extreme_climate.

For citation purposes, cite each article independently as indicated on the article page online and as indicated below:

Lastname, A.A.; Lastname, B.B. Article Title. *Journal Name* **Year**, *Volume Number*, Page Range.

ISBN 978-3-7258-6904-6 (Hbk)
ISBN 978-3-7258-6905-3 (PDF)
https://doi.org/10.3390/books978-3-7258-6905-3

Cover image courtesy of Guoyu Ren

© 2026 by the authors. Articles in this reprint are Open Access and distributed under the Creative Commons Attribution (CC BY) license. The reprint as a whole is distributed by MDPI under the terms and conditions of the Creative Commons Attribution-NonCommercial-NoDerivs (CC BY-NC-ND) license (https://creativecommons.org/licenses/by-nc-nd/4.0/).

Contents

About the Editors

Guoyu Ren

Guoyu Ren, Ph.D., is a Professor and Chief Expert (2004-2020) at the National Climate Center, China Meteorological Administration, and a Discipline Distinguished Scientist (Lead Professor) at the China University of Geosciences (Wuhan) (2015-2025). His research has focused on global, regional, and local climatology and climate change/variability, spanning multi-decade to centennial-scale climate variability, global warming, and urban climate change.

He is a principal investigator on a couple of national key R&D projects/sub-projects, an author/coauthor of 430 articles, and an author/editor of twelve books and seven Special Issues of academic journals. He is among the most cited scientists in atmospheric science and earth sciences, with Google Scholar (GS) citations of more than 27,100 and a GS H-index of 80.

Dr. Ren serves as a member, responsible editor, or associate editor on the editorial committees or boards of 13 international and Chinese journals, and as a member of five specialized committees of the Chinese Meteorological Society and the Chinese Geographical Society. He is a lead expert for the 1st China National Assessment Report on Climate Change (NARCC) and a chapter lead author for the 2nd and 3rd China NARCC. He also serves as a coordinator for the Atmospheric Circulation Reconstructions over the Earth (ACRE) initiative, China regional focus (ACRE China), and as an expert for the World Meteorological Organization Research Board Task Team - Data Exchange with the Research Sector.

Yali Luo

Yali Luo is a professor at the School of Atmospheric Sciences, Nanjing University of Information Science and Technology. Her research focuses on the dynamics and impacts of extreme weather and climate events (convective storms, extreme precipitation, heatwaves), specializing in multi-scale cloud and convective processes analysis. She adopts high-resolution observational analysis and process-oriented modeling, integrating satellite/ground-based observations and advanced numerical simulations, to reveal the governing mechanisms of such events, improve their prediction/projection under global warming, and support resilience planning and risk mitigation. As the PI of an NSFC project and the Co-PI of an NSFC-RGC project, her ongoing research explores Meiyu weather changes in the lower Yangtze River basin and extreme rainfall predictability over the Guangdong–Hong Kong–Macao Greater Bay Area.

Preface

Rapid urbanization is one of the most prominent global changes in the 21st century, characterized by radical modifications of land surfaces, such as the conversion of natural vegetation, farmland, and wetlands into built-up areas, transportation infrastructure, and other anthropogenic landscapes. These profound land modifications not only reshape the urban surface energy balance, hydrological processes, and atmospheric circulation but also significantly and far-reaching affect the occurrence frequency, intensity, and spatial–temporal patterns of urban extreme climate events, including extreme heatwaves, short-duration heavy rainfall, and other hazardous weather phenomena. Against the background of global climate change, the interaction between urbanization-related radical land modification and urban extreme climate has increasingly become the focus of interdisciplinary research in meteorology, geography, urban planning, and environmental science. Understanding the intrinsic mechanisms of this interaction is crucial for improving the accuracy of urban extreme climate prediction, formulating scientific urban adaptation and mitigation strategies, and promoting the sustainable development of cities.

To systematically showcase the latest research progress in this field, we launched this Special Issue (SI), entitled "Influence of Urbanization-Related Radical Land Modification on Urban Extreme Climate," in *Land*. Since the call for papers, this Special Issue has attracted extensive attention and positive responses from researchers worldwide. After a rigorous peer-review process (each submission was reviewed by at least two independent experts in the field), we are pleased to present a collection of eight high-quality research articles that span a wide range of geographical regions, diverse research methods, and in-depth thematic explorations. Collectively, these articles provide valuable insights into the complex relationship between urbanization-driven land modification and urban extreme climate.

Geographically, the research in this Special Issue spans multiple typical urban agglomerations and key cities in China, reflecting East Asia's rapid urbanization. Specifically, the articles focus on Hangzhou, Wuhan, Shanghai, Nanjing, Beijing, Shenyang, and two urban agglomerations in Southeast China. This broad geographical coverage provides a comprehensive understanding of the spatial heterogeneity of the impact of urbanization-related land modification on extreme climate across different climatic zones (e.g., subtropical monsoon climate, temperate monsoon climate) and urban development stages.

Thematically, the collected articles address a series of critical scientific issues.

- Several studies examine the impact of urbanization on extreme temperature events, for example, by exploring the roles of synoptic weather patterns and urbanization in long-term variations in warm and cold events in Nanjing; conducting a comparative analysis of the characteristics and synoptic backgrounds of extreme heat events across two urban agglomerations in Southeast China; and assessing the spatial variability of daytime and nighttime extreme heatwaves in Beijing under different land-use types.
- Other studies focus on the response of extreme precipitation to urbanization, including investigating urbanization-induced diurnal variation in short-duration rainfall events in Wuhan; analyzing how urbanization and the monsoon jointly modify Meiyu precipitation changes in Shanghai; and exploring how urbanization influences the spatiotemporal characteristics of extreme hourly precipitation in Shenyang.
- Additionally, there are studies that adopt a broader perspective, such as assessing the impact of urbanization on extreme climate indices in the entire Yangtze River Economic Belt, and

conducting a sensitivity study on the impacts of urbanization and its parameters on thermal and dynamic fields in Hangzhou using advanced numerical models.

Methodologically, the articles integrate multiple approaches, including in-situ observation data analysis, remote sensing inversion, numerical simulation, and statistical modeling, thereby enriching the research paradigm in this field and enhancing the reliability and depth of the research conclusions.

As Guest Editors of this Special Issue, we would like to express our sincere gratitude to all authors for their valuable contributions of high-quality research. Their efforts have laid the foundation for the success of this Special Issue. We also extend our heartfelt thanks to the Editorial team of *Land* for their professional and efficient support throughout the entire process, from the initial planning of the Special Issue to the final publication. Furthermore, we are deeply grateful to the anonymous reviewers for their rigorous, constructive, and insightful comments, which have significantly improved the quality of the published articles. Finally, we hope that this Special Issue will serve as a platform for academic exchange and collaboration, inspiring more researchers to engage in in-depth studies on the interaction between urbanization-related radical land modification and urban extreme climate. We believe that the findings presented in this Special Issue will provide important scientific references for urban planners, meteorological forecasters, and policymakers to address the challenges posed by urban extreme climate and promote the construction of climate-resilient cities.

Guoyu Ren and Yali Luo
Guest Editors

land

Article

The Impact of Urbanization on Extreme Climate Indices in the Yangtze River Economic Belt, China

Wentao Yang [1,2,*], Yining Yan [1,2], Zhibin Lin [2], Yijiang Zhao [3], Chaokui Li [1,2], Xinchang Zhang [4] and Liang Shan [5,6]

1 National-Local Joint Engineering Laboratory of Geospatial Information Technology, Hunan University of Science and Technology, Xiangtan 411100, China
2 Department of Geographical Information Science, Hunan University of Science and Technology, Xiangtan 411100, China
3 Department of Computer Science, Hunan University of Science and Technology, Xiangtan 411100, China
4 Department of Geographical Information Science, Guangzhou University, Guangzhou 510006, China
5 Guangzhou Zengdian Institute of Science and Technology, Guangzhou 511300, China
6 Department of Business Management, Central South University, Changsha 410083, China
* Correspondence: 8570003@hnust.edu.cn; Tel.: +86-151-1628-0680

Abstract: Urbanization has been proven to be a critical factor in modifying local or regional climate characteristics. This research aims to examine the impact of urbanization on extreme climate indices in the Yangtze River Economic Belt (YREB), China, by using meteorological observation data from 2000 to 2019. Three main steps are involved. First, a clustered threshold method based on remote-sensing nighttime light data is used to extract urban built-up areas, and urban and rural meteorological stations can be identified based on the boundary of urban built-up areas. Nonparametric statistical tests, namely, the Mann–Kendall test and Sen's slope, are then applied to measure the trend characteristics of extreme climate indices. Finally, the urbanization contribution rate is employed to quantify the impact of urbanization on extreme climate indices. The results indicate that urbanization has a more serious impact on extreme temperature indices than on extreme precipitation indices in the YREB. For extreme temperature indices, urbanization generally causes more (less) frequent occurrence of warm (cold) events. The impact of urbanization on different extreme temperature indices has heterogeneous characteristics, including the difference in contamination levels and spatial variation of the impacted cities. For extreme precipitation indices, only a few cities impacted by urbanization are detected, but among these cities, urbanization contributes to increasing the trend of all indices.

Keywords: extreme climate events; YREB; remote-sensing nighttime light data; trend analysis

1. Introduction

The last decades have witnessed the frequent occurrence of extreme climate events across the world. Although a global warming hiatus (pause or slowdown) is a period of relatively little change in globally averaged surface temperatures, which is mainly manifested by the lower warming rate of the global mean air surface temperature when compared with 1998–2012 [1], the 2000s was the warmest decade on record since 1850, with many regional and global records broken. According to the fifth assessment reports published by the Intergovernmental Panel on Climate Change (IPCC), there is little doubt that extreme climate events with unprecedented frequencies, intensities, and durations will continue to occur in the future [2]. The significant damage to human life and the ecological environment as a result of these extreme climate events has severely hindered sustainable development. Consequently, the 2030 agenda of sustainable development goals released by the United Nations clearly states that all countries should adopt urgent action to address climate change and its impacts [3]. Before measurement and policymaking, it is more significant to understand the association mechanism between extreme climate events and the driving factors.

Urbanization has been proven to be one of the critical driving factors for climate change [4,5]. It changes the radiation, thermal and dynamic characteristics of the underlying surface and alters the land cover in an urban area to be completely distinct from that in surrounding areas, which will result in a horizontal gradient of moisture and energy between the urban area and its surrounding areas and further influence the local or regional climate characteristics [6]. Urban climate change, as a result of urbanization, affects the physical environment more directly than global climate change [7]. According to this fact, a series of studies have focused on analyzing the impact of urbanization on climate change. The main differences in existing research are mainly reflected in the analytical object and study area.

As a common analytical object or a representative phenomenon of urban climate change, the urban heat island (UHI) refers to the warmer temperatures experienced by a city than its surroundings [8]. The UHI magnitude or intensity can be evaluated from two perspectives: one is based on the energy balance differences between urban and rural areas, and the other is by comparing the air temperature between urban and rural areas. The former attempts to compute the types and sizes of heat fluxes generated within an area and then apply the energy balance calculation to account for UHI phenomena [9]. The energy balance can be used for clearly understanding the physical mechanism behind the UHI, but it is difficult to accurately obtain the information about heat fluxes. In contrast, although using the difference in air temperature between urban and rural areas cannot explain the physical mechanism, it is easy to determine the UHI intensity because only air temperature data at different sites need to be obtained. Literature statistics show that this method has been commonly applied to quantify UHIs [10,11].

In addition to the effects on air temperatures, urbanization can also affect extreme precipitation events by changing the formation and development of aerosols, clouds, etc. [12,13]. As demonstrated through an increasing number of studies, these extreme precipitation events are associated with increased urbanization [7–9]. Similar to the evaluation of UHI intensity, the commonly used method can also be divided into physical modelling and statistical analysis. Physical modelling aims to apply numerical simulation technology, such as the weather research and forecasting model (WRF), for qualitatively identifying the physical mechanisms of urbanization on precipitation [14–16]. However, the requirement of high-resolution data and the uncertainty of the model parameters are obvious limitations. An alternative method is statistical analysis, which is based on the comparison of precipitation between urban and rural stations [17]. Due to the simple computing process, statistical analysis has also been widely used in existing research. Therefore, the statistical method for evaluating the impact of urbanization on extreme climate events was used in this study.

Nevertheless, whether the statistical method is used to analyze the effects of urbanization on UHI intensity or extreme precipitation, how to select urban and rural stations poses significant challenges because many relevant local-scale aspects need to be considered [18]. According to the difference in the main data source, existing methods for classifying stations can be roughly divided into two categories: one is mainly based on climate data, and the other is based on auxiliary data (population data, land use and land cover (LULC)). The former, such as empirical orthogonal function (EOF) decomposition and interpolation isotherms, attempt to address the classification issue by highlighting different thermal or climate characteristics in urban-rural pairs [19,20]. These methods can exhibit an obvious advantage when considering the boundary of the urbanization effect, but the sparse distribution of meteorological stations in most urban areas cannot effectively support these mathematical methods.

In contrast, these methods based on auxiliary data first extract urban built-up areas, and meteorological stations located in (outside) the urban built-up areas are then identified as urban stations (rural stations). Obviously, the crucial issue is to accurately obtain urban built-up areas. The spatial distribution of population density is one of the main sources used to extract urban built-up areas. Specifically, a threshold for the population needs to be set in advance, and an area with a higher value than the threshold is identified as a built-up area [21]. Nevertheless, it is difficult to determine the threshold because the population change with urban areas and to obtain accurate population data at a fine scale need to be further discussed. Remote sensing nighttime light data are useful for extracting large-scale built-up urban areas and have also been used to identify urban and rural stations [22]. Compared with other auxiliary data, namely, population data and LULC, remote sensing nighttime light data can be directly collected by sensors and effectively reveal the area of human activities, which has attracted increasing attention in extracting urban built-up areas. However, empirical threshold-based methods were selected to extract urban built-up areas in previous studies on meteorological station classification [23]. They may overestimate the built-up areas in urban regions due to the 'blooming' effect of nighttime lights and over small patches in developing towns because of the relatively low value of nighttime lights. Consequently, an inaccurate result of urban built-up areas will result in error in station classification and further lead to an unreliable result of the impact of urbanization on UHIs or extreme precipitation. In this research, a clustered threshold strategy was used to extract urban built-up areas by remote sensing nighttime light data, which can avoid the above issues resulting from the empirical threshold-based method.

In addition, due to the spatial heterogeneity of urban characteristics, such as land cover, impermeable layers, and industrial structures, the impact of urbanization on UHIs or extreme climate indices may vary with the study area or scale [24–27]. The Yangtze River Economic Belt (YREB), China, was selected as the study area. The YREB was positioned as a coordinated development belt for western, central, and eastern parts of China, a pioneering and demonstration belt for ecological and green civilization, and an inland river economic belt with a global influence [28]. The outline of the Yangtze River Economic Belt Development Plan, released in March 2016, further emphasized a development pattern for the area with concentrations on ecological and green civilization [29]. Undoubtedly, this national strategy would provide an unprecedented opportunity for in-depth development of the cities in the YREB. However, in recent decades, because of urbanization processes, climate conditions in the YREB have obviously changed [30]. The disastrous floods and extreme temperature events associated with climate change occurred frequently and resulted in considerable economic loss and human lives in the YREB [31]. Therefore, it is of great significance to understand the impact of urbanization on extreme climate events in the YREB.

The remaining paper is organized as follows. Section 2.1 introduces the study area and datasets. Section 2.2 introduces the method for identifying the impact of urbanization on extreme climate indices. The experimental results are described in Section 3. The conclusion and discussion are presented in Sections 4 and 5, respectively.

2. Materials and Methods

2.1. Study Area and Datasets

As the longest river in China and the third-longest river in the world, the mainstream Yangtze River flows through two municipalities (Chongqing and Shanghai) and nine provinces (Anhui, Guizhou, Hubei, Hunan, Jiangsu, Jiangxi, Sichuan, Yunan, and Zhejiang). The area covering these 11 provincial-level regions is called the YREB, which occupies 2.05 million square kilometers, accounting for approximately 21.3% of China's land area. In 2017, the total population and GDP in the area were approximately 595 million and 37,100 billion yuan, which correspond to 42.8% and 44.9% of the country's total, respectively [29]. The spatial distribution of the 11 provincial-level regions is shown in Figure 1, and each region is marked with a unique colour.

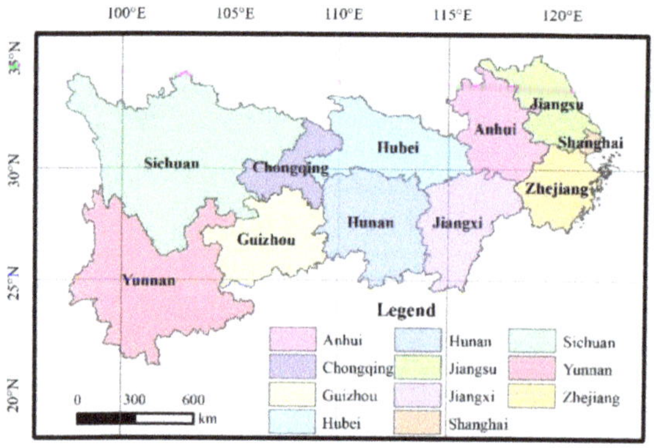

Figure 1. Spatial distribution of the 11 provincial-level regions.

The time series of daily maximum/minimum temperature and daily precipitation for 20 years (2000–2019) were obtained from the China Meteorological Data Service Centre (http://data.cma.cn/en accessed on May 12, 2022). After the meteorological stations with a large amount of missing data were removed (the proportion of missing data at some sites was higher than 30%), there were 252 meteorological stations in the YREB, accounting for 96% of the total number of all stations. Based on meteorological data at 252 stations, the software package RClimDex (http://etccdi.pacificclimate.org/software.shtml accessed on May 12, 2022) was applied to compute 27 extreme climate indices, including 16 extreme temperature indices and 11 extreme precipitation indices [32]. A brief introduction of these extreme climate indices is shown in Table 1. In addition, the vector map in the YREB [33] was also collected as auxiliary data for visualization and verification analysis. To distinguish urban and rural meteorological stations, remote-sensing nighttime light data [34] and built-up area data [35] were collected.

Table 1. Extreme climate indices.

Category	Index	Unit	Definition
Extreme temperature indices	FD0 *	day	Number of days with daily minimum temperature (TN) < 0°C
	ID0 *	day	Number of days with daily maximum temperature (TX) < 0°C
	TN10P *, TN90P #	day	Number of days with TN < 10th (>90th) percentile
	TX10P *, TX90P #	day	Number of days with TX < 10th (>90th) percentile
	TR20 #, SU25 #	day	Number of days with TN > 20 °C (TX > 25 °C)
	TNn, TNx/TXn, TXx	°C	Minimum (maximum) value of TN/TX
	DTR	K	Difference between TX and TN
	CSDI, WSDI	day	Number of days with TN < 10th (>90th) percentile for at least 6 consecutive days
	GSL	day	Numbers of days with daily average temperature>5°C
Extreme precipitation indices	SDII	mm/day	Annual precipitation divided by the number of wet days
	CDD	day	Maximum number of consecutive days with daily precipitation amount (RR) < 1 mm (Maximum length of dry spell)
	CWD	day	Maximum number of consecutive days with RR ≥1 mm
	RX1day	mm	Maximum daily precipitation amount
	RX5day	mm	Maximum precipitation amount for 5 consecutive days
	R10mm, R20mm, R25mm	day	Number of days with RR > 10mm (20, 25 mm)
	R95P, R99P	day	Number of days with RR > 95th (99th) percentile
	Prcptot	mm	Annual precipitation amount

Warm indices: TN90P, TX90P, TR20, and Su25; * Cool indices: FD0, ID0, Ta10P, TN10P, and TX10P.

Admittedly, because of changes in observation stations, instruments, schedules, and habits, inhomogeneous data or discontinuous points can be created [36]. The inhomogeneous data may cause deviations in estimating climate trends, resulting in inaccurate analyses for regional climate detection [37]. It is critical to perform data homogenization so that the homogenous time series of climate variables contain only climate variation and regional trend information. In this study, there are nine stations whose locations have changed at least once. We deleted these climate data from these stations instead of adjusting them. Compared with the number of all stations, the number of stations with location changes is less, and detecting these few stations may not have much impact on the analysis results.

2.2. Methods

As mentioned in the first section, the statistical method for evaluating the impact of urbanization on extreme climate events was used in this study [38–40], but how to accurately identify urban and rural stations is crucial for the statistical method to evaluate urbanization on extreme climate indices. This research aims to apply a remote sensing method of selecting urban and rural stations for evaluating urbanization on extreme climate indices in the Yangtze River Economic Belt, China. The framework of this research is shown in Figure 2. The main process can be divided into three steps. First, a clustered threshold strategy based on remote-sensing nighttime light data is used to extract urban built-up areas [41], and urban and rural stations can be identified by the boundary of urban built-up areas [42–44], as described in Section 2.2.1. Second, the Mann–Kendall test is applied to determine the significant trends among the time series of extreme climate indices [45]. For the time series of an index at one station, if a significant increasing or decreasing trend is confirmed, then Sen's slope method is used to estimate the magnitude of the change trends of extreme climate indices [25]. Otherwise, the time series at the station is ignored in the subsequent analysis. The detailed introduction is described in Section 2.2.2. Third, based on the change trend of extreme climate indices at urban and rural stations, the urbanization contribution rate is applied to assess the impact of urbanization on extreme climate indices, as described in Section 2.2.3.

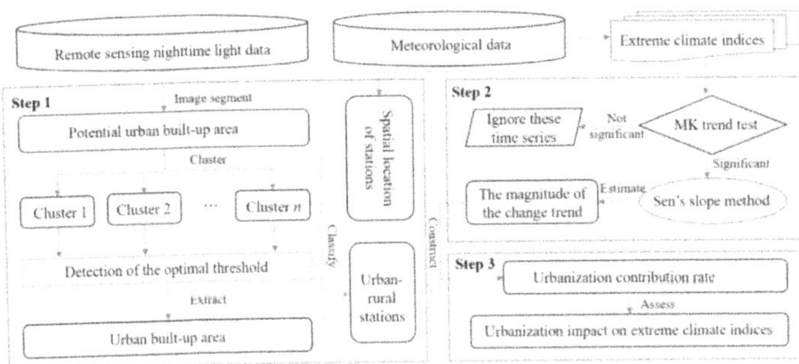

Figure 2. Framework of the statistical method for analyzing the impact of urbanization on extreme climate indices.

2.2.1. Classifying Meteorological Stations Using a Remote Sensing Method

The remote sensing method was used in this research to classify meteorological stations into urban and rural stations. This method first extracts urban built-up areas by using remote sensing nighttime light data, and according to the topological relation between urban built-up areas and meteorological stations, the meteorological stations will be classified into urban and rural stations. Specifically, the meteorological stations located in the urban built-up areas are defined as urban stations, and those located outside the urban built-up areas are defined as rural stations. Obvious differences in nighttime light intensity exist among urban areas with different urbanization levels because of varied urban development stages and energy consumption patterns. Therefore, the threshold for extracting urban built-up areas from nighttime light should change with the level of nighttime light intensity. It is reasonable to apply a clustered threshold strategy to handle this task.

The clustered threshold strategy mainly includes three steps. First, an image segmentation algorithm is used to obtain the potential urban built-up areas, which is defined as a set of areas composed of spatially continuous lighted pixels in the single-band nighttime light. A series of image segmentation algorithms have been proposed for object-based classification in remote sensing. Considering that the marker-controlled watershed segmentation algorithm based on greyscale morphology is suitable for single-band remote sensing nighttime light images, this algorithm is used to obtain the potential urban built-up areas. The detailed process can be found in the relevant literature [41].

Then, a clustering algorithm is applied to divide the potential urban built-up areas into several groups with similar nighttime light intensities. As only the total digital number of nighttime lights needs to be considered in the clustering process, the classical k-means algorithm is used to divide the potential urban built-up areas [46]. The elbow method was used to determine the optimal cluster number k [47]. It considers the variation characteristics of the sum of squared deviations from the means in each cluster, the within sum of squares (WSS), which is defined as

$$WSS(k) = \sum_{j=1}^{k} \sum_{i=1}^{N_j} (x_{ij} - \overline{x}_j)^2 \tag{1}$$

where N_j is the number of objects belonging to the jth cluster; x_{ij} and $\overline{x}_j$ are the total DN values of the ith objects of the jth cluster; and $\overline{x}_j$ is the mean DN value of all the objects of the jth cluster. At some value for k, the WSS drops dramatically, and after that it reaches a plateau when the value of k continues to increase. This point, when this decrease in WSS dispersion slows down, is called the "elbow" [48]. In this situation, the optimal cluster k can be identified, and the clustering result is used to determine different thresholds for extracting urban built-up areas.

Finally, for each cluster, the optimal threshold to classify the urban and rural areas is detected by comparing the areas of the statistical data and urban built-up areas with different thresholds. That is, the statistical data through an official survey always accurately record the total area of each urban built-up area. The different results of the built-up areas can be obtained based on a set of thresholds. The total area of the built-up area can be directly based on the number of pixels whose DN values are larger than the threshold. The optimal threshold corresponds to the built-up area closest to the statistical data. Specifically, assume that the set of thresholds is $\boldsymbol{t} = \{t_1, \ldots, t_n\}$ ($i = 1, \ldots, n$), and for each threshold, the total area of the built-up areas $A(t_i)$ ($i = 1, \ldots, n$) can be obtained by counting the number of pixels whose DN values are larger than t_i. The optimal threshold t' for a cluster can be computed as

$$t' = argmin\, |A(t_i) - S| \tag{2}$$

where S is the total area of each urban built-up area from an official survey.

2.2.2. Identifying the Change Trends of Extreme Indices

In this study, the nonparametric Mann–Kendall (MK) test is used to detect the significant trends among the time series of extreme climate indices. A positive (negative) value of the MK statistic indicates an increasing (decreasing) trend in the data. For a sample size >10, a normal approximation to the Mann–Kendall statistics Z can be applied to test significance. If Z is greater than $Z_{\alpha/2}$, where α is the significance level (0.10 in the study), it is considered a significant trend [49].

The Sen's slope method was selected to estimate the magnitude of the change trend of extreme climate indices. As a nonparametric method, Sen's slope assumes a linear trend in the time series and then applies a linear model to calculate the slope of the trend [50]. A positive value of the estimator corresponds to an increasing trend, and a negative value indicates a decreasing trend in the time series. As the slope estimator is calculated through the median statistic, Sen's slope is robust to outliers and missing data.

2.2.3. Assessing the Impact of Urbanization on Extreme Climate Indices

The impact of urbanization on extreme climate indices can be reflected by the linear trend changes in extreme climate indices at urban stations resulting from urbanization [51]. The impact value Δβ can be defined as:

$$\Delta\beta = \beta_u - \beta_r \quad (3)$$

where β_u and β_r indicates the change trend of extreme climate indices at urban and rural stations, respectively. A positive value of the impact value $\Delta\beta$ corresponds to an increasing trend of urbanization on extreme climate indices, and a negative value indicates a decreasing trend of urbanization on extreme climate indices. Furthermore, the urbanization contribution rate is expressed as:

$$r_u = |\Delta\beta/\beta_u| \times 100\% \quad (4)$$

where if r_u is equal to 0, it means that urbanization has no contribution to the extreme index extreme at urban stations, and if r_u is equal to 100%, it means that the change trend of extreme climate indices at urban stations is completely caused by urbanization. Notably, because some unknown local factors may exist, the values of the urbanization contribution rate may be greater than 100%, and in this case, the urbanization contribution rate is also regarded as 100% [52].

3. Results

3.1. Classification of Meteorological Stations

As introduced in Section 2.2.1, the watershed segmentation algorithm was first used to obtain the potential urban clusters. The spatial distribution of potential urban built-up areas is shown in Figure 3, and the orange area indicates the potential urban built-up area. The nighttime light intensity of potential urban built-up areas was then aggregated into prefecture-level cities based on the administrative boundary. Furthermore, the number of clusters defined using the k-means cluster algorithm over the nighttime light intensity was tested for k = 2, . . . , 8. The estimates of WSS for different k are visualized in Figure 4. The number of clusters was set to 4 since the WSS drops dramatically when k is from 3 to 4 and reaches a plateau when k is from 4 to 5. The cluster result is presented in Figure 5, and these four clusters are displayed in different colors. From Cluster 1 to 4, the nighttime light intensity of prefecture-level cities gradually increases, but the number of these cities gradually decrease.

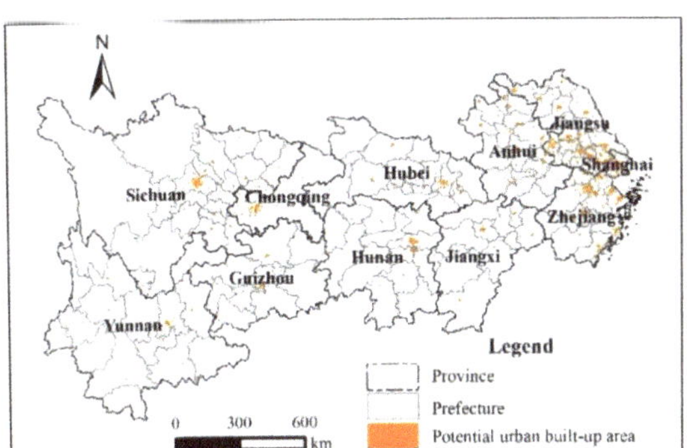

Figure 3. Spatial distribution of potential urban built-up areas.

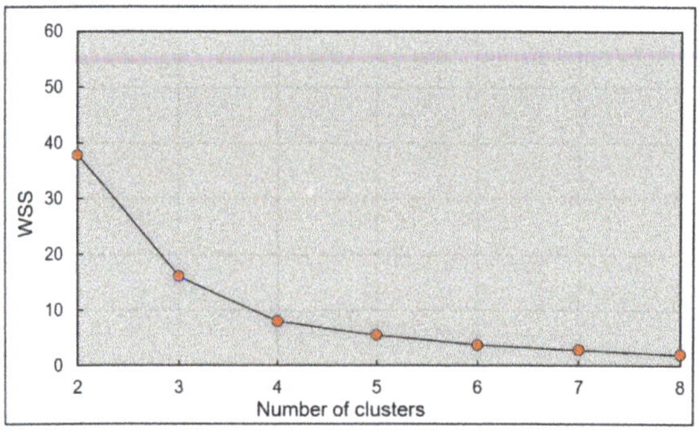

Figure 4. The estimates of WSS for different k.

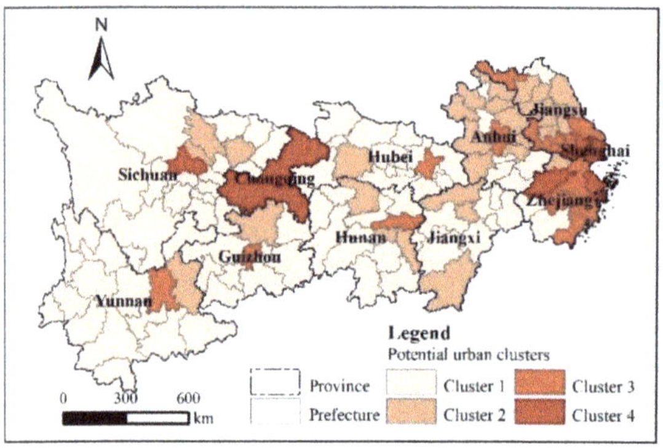

Figure 5. Cluster results of potential urban clusters.

For each cluster, different results of extracting the built-up areas can be computed based on a set of thresholds. In Figure 6, the horizontal axis indicates the DN threshold, and the vertical axis represents the absolute value of the difference between the built-up area from the remote sensing data and the statistical data. Based on Equation (2), the optimal threshold corresponds to the lowest value of the cures in Figure 6. Therefore, the optimal thresholds for Cluster 1 to 4 are 2, 18, 21, and 23, respectively. According to the optimal threshold of each cluster, the final urban built-up areas can be identified. Furthermore, urban and rural stations can be separated based on the boundary of these built-up areas. The total number of urban and rural stations are 203 and 41, respectively. The spatial distribution of urban and rural stations is shown in Figure 7, and the orange squares and blue circles indicate rural and urban stations, respectively.

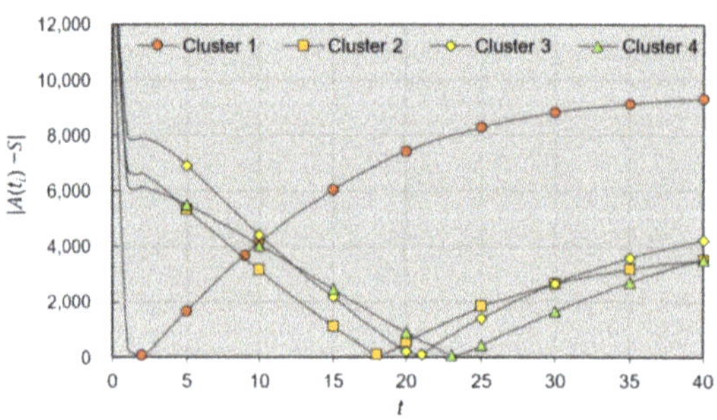

Figure 6. The estimates of WSS for different k.

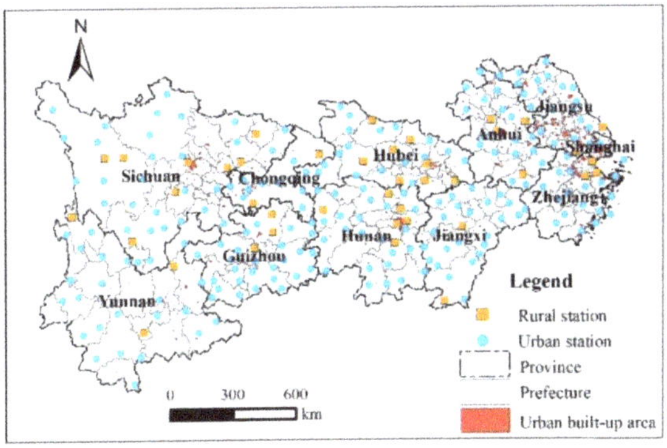

Figure 7. Spatial distribution of urban and rural stations.

Notably, a potential assumption in Equation (3) is that the urban and rural stations are usually distributed in adjacent locations. Due to the significant difference in the number of stations between urban and rural areas, the nearest rural stations of some urban stations may be far away from them and have different climate characteristics that not only result from the urbanization. In that case, the urban stations should be ignored because of the lack of nearby rural stations as a reference. Consequently, to filter out these cities and retain the prefecture-level cities with higher urbanization levels, we built a 110 km buffer zone centered on each rural station, and these urban stations within a buffer are reserved. Here, a buffer radius that is too large violates the principle of proximity between urban and suburban stations, which may lead to extreme climate impacts not completely caused by urbanization. Too small a buffer radius may lead to many cities not referring to rural stations. Considering the above conditions, the buffer radius was selected as 110 km. In this situation, the radius was not too large, and more rural stations could be matched to urban stations. Finally, 99 urban stations (or cities) and 38 rural stations were determined, which are shown in Figure 8.

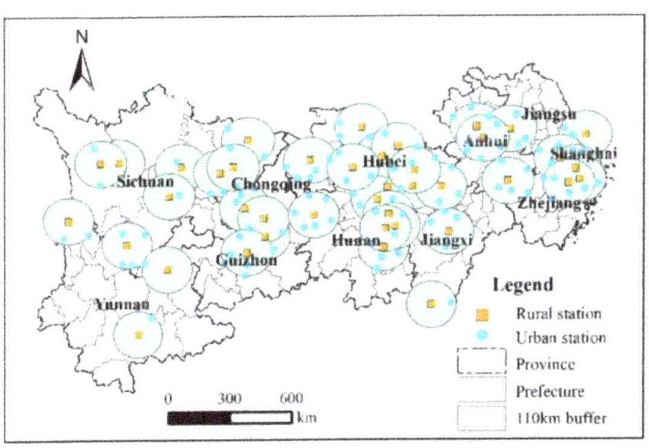

Figure 8. Result of matching urban and rural stations based on 110 km buffer.

3.2. The impact of Urbanization on Extreme Temperature Indices

For each extreme temperature indices, the MK test and Sen's slope method were first used to detect the significant trends among time series at both urban and rural stations. The statistical results are listed in Figure 9. The horizontal axis represents extreme temperature indices, and the vertical axis indicates the proportion of cities with different trend types in the number of all cities. Overall, these indices, including TN90P, TR20, TNx, and TX90P, exhibit significant upward trends, and TX10P and TN10P show significant downwards trends at most stations. There are no significant or dominant trends in the remaining indices. The indices with significant upward trends mainly consist of warm indices (TN90P, TR20, and TX90P). Meanwhile, all the indices with significant downwards trends (TN10P, and TX10P) are cold indices.

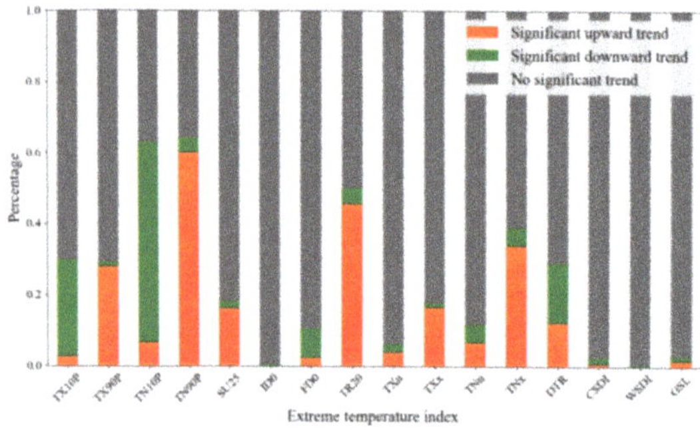

Figure 9. Percentage of meteorological stations with different trends for the extreme temperature index.

Furthermore, based on Equations (3) and (4), the urbanization contribution rates on TN90P, TR20, TNx, TX90P, TN10P, and TX10P can be obtained. The spatial distribution and urbanization contribution rates of the impacted cities are shown in Figure 10. Only cities in which extreme temperature indices had a significant impact caused by urbanization are labelled in Figure 10. The red and green triangles represent the impact of urbanization with a dominant upwards and downwards trend, respectively, and the size of triangles represents urbanization contribution rates, which was divided into four intervals, namely, [0%, 25%), [25%, 50%), [50%, 75%), and [75%, 100%]. Among all 99 cities, the total number of cities in which TN90P, TR20, TNx, TX90P, TN10P, and TX10P were impacted by urbanization was 36, 24, 9, 11, 20, and 7, respectively. Among these indices affected by urbanization with a dominant upwards trend, TN90P was the most severely affected by urbanization, followed by TR20. Both TNx and TX90P are similarly affected by urbanization, but only a very small number of cities were detected. It can also be found that the spatial distribution of these cities is inconsistent. Cities of which TN90P, shown in Figure 10a, was impacted by urbanization covered most areas of the YERB. However, most of them are distributed in the middle reaches, including Chonqing, Guizhou, Hubei, and Hunan Provinces, where the urbanization contribution rates are mainly located in the range of 25% to 75%. Overall, other cities located upstream and downstream have low contribution rates. In Figure 10b, cities of which TR20 was impacted by urbanization are also mainly distributed in the middle reaches with a high variance of urbanization contribution rates (the range is from 0% to 75%). The downstream cities with high values of urbanization contribution rates are in Anhui Province. For the cities in which TNx and TX90P were impacted by urbanization, as shown in Figure 10c,d, there was no spatial aggregation pattern.

In the indices impacted by urbanization with a dominant downwards trend, TN10P is more seriously impacted by urbanization than TX10P. As shown in Figure 10e, cities corresponding to TN10P are mainly distributed in the middle reaches of the YERB, including Chongqing and Hubei provinces, and many values of urbanization contribution rates are greater than 50%. Although TX10P was not severely impacted by urbanization, as shown in Figure 10f, the spatial distribution of the impacted cities exhibits an agglomeration pattern, and most of the cities are located in Sichuan Province.

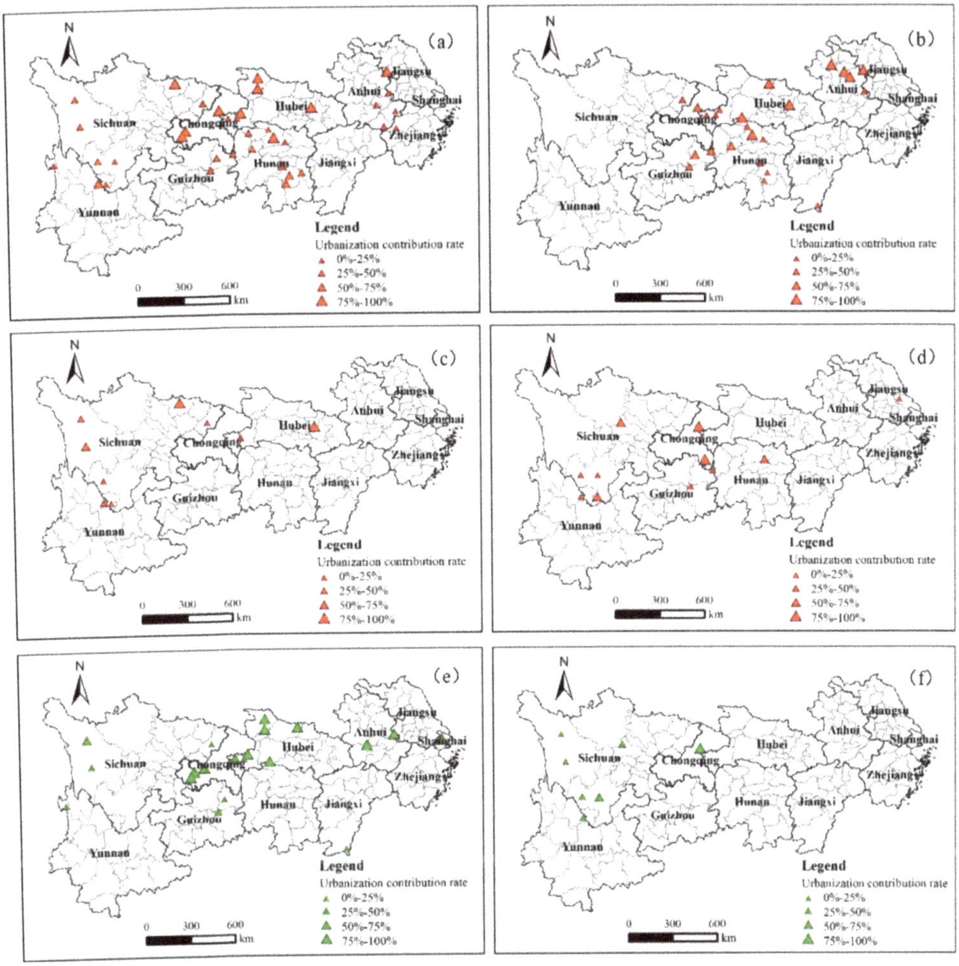

Figure 10. Urbanization contribution rates of different extreme temperature indices: (**a**) TN90P, (**b**) TR20, (**c**) TNx, (**d**) TX90P, (**e**) TN10P, and (**f**) TX10P.

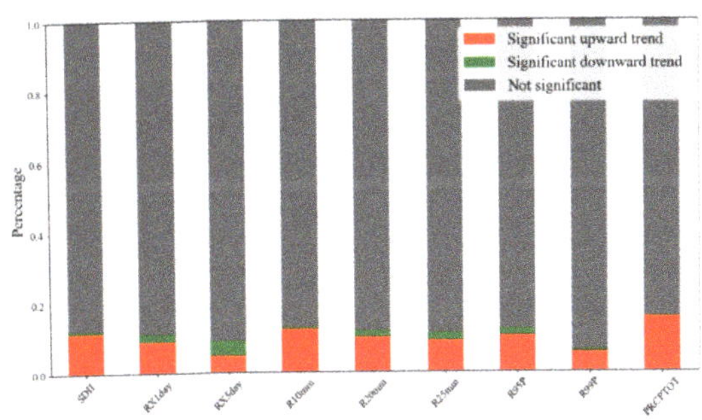

Figure 11. Percentage of meteorological stations with different trends for the extreme precipitation index.

3.3. The Impact of Urbanization on Extreme Precipitation Indices

Similarly, the nonparametric statistical test was applied to identify different types of trends among the time series of extreme precipitation indices at both urban and rural stations. The statistical results are shown in Figure 11. For all extreme precipitation indices, the number of cities with no significant trend is the largest. In the cities in which the index has a significant trend, there are more cities with an upwards trend than those with a downwards trend. Only cities of which Prcptot was impacted by urbanization are spatially clustered and mainly located downstream of YERB, as shown in Figure 12.

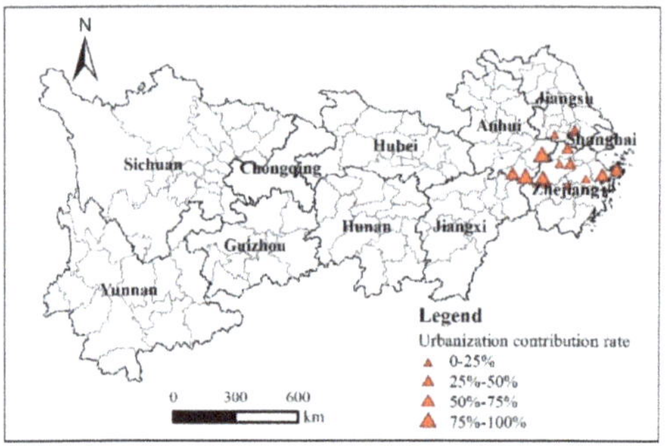

Figure 12. Urbanization contribution rates of the extreme precipitation index Prcptot.

4. Discussion

Understanding the impact of urbanization on extreme climate events is a crucial topic in climate change. This research used the statistical comparison of extreme climate between urban and rural stations to measure the impact of urbanization on extreme climate events. Compared with physical modelling, statistical analysis exhibits obvious advantages and limitations. It can be directly used to measure the impact based only on climate data from different stations, but the method cannot describe the physical mechanism behind these climate processes. Among the statistical methods, an important process was how to classify meteorological stations into urban and rural stations, which was further converted to the issue of the identification of urban built-up areas. Therefore, a clustered threshold strategy based on nighttime light data was then presented to handle this issue. Compared with other methods, such as empirical orthogonal function decomposition and interpolation isotherms, the remote sensing method only involves nighttime light data, which are easily accessible, and the principle for meteorological station classification is also easy to understand.

The YREB with rapid urbanization development was selected as the study area. Previous studies on the extreme climate of the YREB [53] mainly focused on the characteristics of extreme temperatures [45] and extreme precipitation [54] and rarely explored the impact of urbanization on the extreme climate of the YREB [55]. In this research, it was found that the number of meteorological stations with significant trends for extreme temperature indices was higher than that for extreme precipitation indices. This conclusion was also demonstrated in related studies. For example, Cheng et al. found that the number of extreme temperature indices with significant changes in the Yangtze River Basin from 1958 to 2017 was 2.8 times that of the extreme precipitation index [55]. Among these extreme temperature indices, warm indices, including TN90P, TR20, and TX90P, always exhibit an upwards trend, and most of the cold indices, such as TN10P and TX10, show a decreasing trend in the study area, which is also consistent with the results in other regions or local regions in the YREB. For example, Zhao et al. found that the warm indices showed upwards trends in the Beijing-Tianjin-Hebei regions during the period 1980-2015, while the cold indices tended to decrease [25]. Shi et al. found that the main cold index of the extreme temperature index in the Yangtze River Basin showed a downwards trend from 1970 to

2014, and the main warm index of the extreme temperature index showed an upwards trend [45].

The urbanization contribution rates were used to quantitatively measure the impact of urbanization on extreme climate indices. The impact of urbanization on extreme temperature indices has obviously heterogeneous characteristics, which is mainly reflected in the following two aspects. One is that urbanization has distinct levels of impact on different extreme temperature indices, and TN90P, TR20, and TN10P were seriously affected by urbanization. The other is that the impact of urbanization on different indices has spatial variability; that is, for different extreme temperature indices, the spatial distribution of the impacted cities is inconsistent. For example, cities in which TN90P, TR20, and TN10P were impacted by urbanization are mainly distributed in the middle reaches, and most of the cities in which TX10P exhibits a downwards trend are located in Sichuan Province. This similar conclusion was also demonstrated in related studies. For example, Qiu et al. found that extreme temperature indices (TNn, TXx) in China increased significantly from 1960 to 2016 with varying degrees in different seasons and different regions [52]. Lin et al. found that in 20 urban agglomerations in China from 1971 to 2014, for both extreme temperature and precipitation indices, urban and rural areas exhibit remarkably distinct changes and demonstrate a significant urbanization effect, which also varies across different climate backgrounds [26].

The impact of urbanization on extreme precipitation indices was lower than that on extreme temperature indices, which indicates that extreme precipitation indices were not susceptible to urbanization in the YREB. The impacts on extreme precipitation indices display stronger regional discrepancies than those on temperature extremes [26]. In previous studies, there was no consistent conclusion on the impact of urbanization on extreme precipitation indices. Some researchers found that urbanization caused great precipitation and heavy precipitation frequency [56], and one potential reason is that because of the effect of urban heat islands, local thermodynamics and unstable atmospheres may result in stronger precipitation in urban areas [57]. Other researchers demonstrated that the increase in aerosols and the associated cloud microphysical process could cause the reduction in precipitation in the urbanization process [58], and a downwards trend of extreme precipitation indices might be exhibited in some cities [59]. The results in this research demonstrated that the extreme precipitation indices of only a few cities exhibited a significant trend, so urbanization did not absolutely affect the 9 extreme precipitation frequencies.

In addition, we also found that the cities of which Prcptot was significantly impacted by urbanization were concentrated in the Yangtze River Delta, especially in Zhejiang Province. Meanwhile, Prcptot shows a downwards trend in these areas. However, in the study of Yuan et al., the change trend of Prcptot in the Yangtze River Basin from 1961 to 2020 was not significant [50], and Prcptot was not proven to be impacted by urbanization. This research found that TX90P had the most significant growth trend of 5.7 days/10 years only in Yunnan Province. However, Shi et al. found that from 1970 to 2014, warm days (TX90P) in the Yangtze River Basin obviously increased, with a trend of 4.73 days/10 years [45].

This research only focuses on quantifying the impact of urbanization of cities on different extreme climate indices in the YREB and was not conducive to a deep discussion regarding the potential association relationship, namely, how to explain the heterogeneous characteristics of the impact of urbanization on extreme climate indices. The above content needs to be discussed in future research.

5. Conclusions

In this research, the impact of urbanization on extreme temperature and precipitation indices in the YREB was examined by comparing the trend characteristics of urban and rural stations. The specific process mainly includes three steps: a clustered threshold method based on remote-sensing nighttime light data is first applied to extract urban built-up areas, and urban and rural meteorological stations can be identified based on the boundary of urban built-up areas. The total numbers of urban and rural stations were 203 and 41, respectively, and 99 urban stations with corresponding rural stations were finally selected. The nonparametric statistical method is then applied to measure the trend characteristics of extreme temperature and precipitation indices. The warm indices (TN90P, TR20, and TX90P) show significant upwards trends, and all the indices with significant downwards trends (TN10P and TX10P) were cold indices. Except for a few cities in which extreme precipitation indices show an upwards trend, those indices did not exhibit a significant trend in most urban areas. Based on the change trend of extreme climate indices at urban and rural stations, the urbanization contribution rate is applied to assess the impact of urbanization on extreme climate indices. Overall, urbanization has a more significant impact on extreme temperature indices than on extreme precipitation indices. Urbanization leads to a more (less) frequent occurrence of warm (cold) events. The impact of urbanization on different extreme temperature indices has heterogeneous characteristics, which are reflected in the difference in contribution levels and spatial variation of the impacted cities. The extreme precipitation indices of only a few cities exhibited a significant trend, so urbanization did not absolutely affect the frequency of the nine extreme precipitation events.

Author Contributions: Conceptualization, Z.L. and W.Y.; methodology, W.Y., Y.Y, and Y.Z.; software, Y.Z. and L.S.; validation, Y.Y., Z.L., and L.S.; formal analysis, C.L. and X.Z.; investigation, Z.L.; resources, Y.Z.; data curation, Z.L.; writing—original draft preparation, W.Y.; writing—review and editing, W.Y., Y.Y.; visualization, Y.Y., Z.L. and Y.Z.; supervision, C.L. and X.Z.; project administration, W.Y. and X.Z.; funding acquisition, W.Y. All authors have read and agreed to the published version of the manuscript.

Funding: This research was funded by the Foundation for Innovative Research Groups of the Natural Science Foundation of Hunan Province, grant number 2020JJ1003, Natural Science Foundation of Hunan, China, grant number 2021JJ30245 and 2022JJ60015, the Philosophy and Social Science Foundation of Hunan Province, China, grant number 18YBQ050, and Open Research Fund Program of LIESMARS, grant number 19I04.

Institutional Review Board Statement: Not applicable.

Informed Consent Statement: Not applicable.

Data Availability Statement: Not applicable.

Acknowledgments: We express our sincere appreciation to the anonymous reviewer for constructive comments.

Conflicts of Interest: The authors declare no conflict of interest. The funders had no role in the design of the study; in the collection, analyses, or interpretation of data; in the writing of the manuscript, or in the decision to publish the results.

References

1. Chatzopoulos, T.; Perez Dominguez, I.; Zampieri, M.; Toreti, A. Climate extremes and agricultural commodity markets: A global economic analysis of regionally simulated events. *Weather Clim. Extrem.* **2020**, *27*, 100193. [CrossRef]
2. Liao, H.; Chang, W. Integrated assessment of air quality and climate change for policy-making: Highlights of IPCC AR5 and research challenges. *Natl. Sci. Rev.* **2014**, *1*, 176–179. [CrossRef]
3. United Nations. Transforming Our World: The 2030 Agenda for Sustainable Development. Available online: https://sustainabledevelopment.un.org/post2015/transformingourworld/publication (accessed on 12 May 2022).
4. Chen, M.; Xian, Y.; Wang, P.; Ding, Z. Climate change and multi-dimensional sustainable urbanization. *J. Geogr. Sci.* **2021**, *31*, 1328–1348. [CrossRef]
5. Jin, M.; Dickinson, R.E.; Zhang, D.A. The footprint of urban areas on global climate as characterized by Modis. *J. Clim.* **2005**, *18*, 1551–1565. [CrossRef]
6. Zhang, N.; Gao, Z.Q.; Wang, X.M.; Chen, Y. Modeling the impact of urbanization on the local and regional climate in Yangtze River Delta, China. *Theor. Appl. Climatol.* **2010**, *102*, 331–342. [CrossRef]
7. Tam, B.Y.; Gough, W.A.; Mohsin, T. The impact of urbanization and the urban heat island effect on day to day temperature variation. *Urban Clim.* **2015**, *12*, 1–10. [CrossRef]
8. Bassani, F.; Garbero, V.; Poggi, D.; Ridolfi, L.; von Hardenberg, J.; Milelli, M. An innovative approach to select urban-rural sites for Urban Heat Island analysis: The case of Turin (Italy). *Urban Clim.* **2022**, *42*, 101099. [CrossRef]

9. Oke, T.R.; Mills, G.; Christen, A.; Voogt, J.A. *Urban Climates*; Cambridge University Press: Cambridge, UK, 2017.
10. Stewart, I.D. A systematic review and scientific critique of methodology in modern urban heat island literature. *Int. J. Climatol.* **2011**, *31*, 200–217. [CrossRef]
11. Kim, S.W.; Brown, R.D. Urban heat island (UHI) intensity and magnitude estimations: A systematic literature review. *Sci. Total Environ.* **2021**, *779*, 146389. [CrossRef]
12. Shepherd, J.M.; Carter, M.; Manyin, M.; Messen, D.; Burian, S. The impact of urbanization on current and future coastal precipitation: A case study for Houston. *Environ. Plan. B Plan. Des.* **2010**, *37*, 284–304. [CrossRef]
13. Yang, P.; Ren, G.; Yan, P. Evidence for a strong association of short-duration intense rainfall with urbanization in the Beijing urban area. *J. Clim.* **2017**, *30*, 5851–5870. [CrossRef]
14. Yang, L.; Smith, J.; Niyogi, D. Urban impacts on extreme monsoon rainfall and flooding in complex terrain. *Geophys. Res. Lett.* **2019**, *46*, 5918–5927. [CrossRef]
15. Li, Y.; Wang, W.; Chang, M.; Wang, X. Impacts of urbanization on extreme precipitation in the Guangdong-Hong Kong-Macau Greater Bay Area. *Urban Clim.* **2021**, *38*, 100904. [CrossRef]
16. Hu, Z.; Li, Y. Effect of urbanization on extreme temperature events in Liaoning Province, China, from a spatiotemporal perspective. *Urban Clim.* **2022**, *41*, 101025. [CrossRef]
17. Kang, C.; Luo, Z.; Zong, W.; Hua, J. Impacts of Urbanization on Variations of Extreme Precipitation over the Yangtze River Delta. *Water* **2021**, *13*, 150. [CrossRef]
18. Stewart, I.D. Landscape representation and the urban-rural dichotomy in empirical urban heat island literature, 1950–2006. *Acta Climatol. Chorol.* **2007**, *40*, 111–121.
19. Wu, F.T.; Fu, C.; Qian, Y.; Gao, Y.; Wang, S.Y. High-frequency daily temperature variability in China and its relationship to large-scale circulation. *Int. J. Climatol.* **2017**, *37*, 570–582. [CrossRef]
20. Anderson, C.I.; Gough, W.A.; Mohsin, T. Characterization of the urban heat island at Toronto: Revisiting the choice of rural sites using a measure of day-to-day variation. *Urban Clim.* **2018**, *25*, 187–195. [CrossRef]
21. Ren, Y.; Ren, G. A Remote-Sensing Method of Selecting Reference Stations for Evaluating Urbanization Effect on Surface Air Temperature Trends. *J. Clim.* **2011**, *24*, 3179–3189. [CrossRef]
22. Owen, T. Using DMSP-OLS light frequency data to categorize urban environments associated with US climate observing stations. *Int. J. Remote Sens.* **1998**, *19*, 3451–3456. [CrossRef]
23. Su, Y.; Chen, X.; Wang, C.; Zhang, H.; Liao, J.; Ye, Y.; Wang, C. A new method for extracting built-up urban areas using DMSP-OLS nighttime stable lights: A case study in the Pearl River Delta, southern China. *GISci. Remote Sens.* **2015**, *52*, 218–238. [CrossRef]
24. Park, J.; Kim, M.-K. Correlation between Urbanization Rate in Local Scale and Extreme Climate Indices. *J. Clim. Res.* **2013**, *8*, 185–201. [CrossRef]
25. Zhao, N.; Jiao, Y.; Ma, T.; Zhao, M.; Fan, Z.; Yin, X.; Liu, Y.; Yue, T. Estimating the effect of urbanization on extreme climate events in the Beijing-Tianjin-Hebei region, China. *Sci Total Environ.* **2019**, *688*, 1005–1015. [CrossRef] [PubMed]
26. Lin, L.; Gao, T.; Luo, M.; Ge, E.; Yang, Y.; Liu, Z.; Zhao, Y.; Ning, G. Contribution of urbanization to the changes in extreme climate events in urban agglomerations across China. *Sci. Total Environ.* **2020**, *744*, 140264. [CrossRef] [PubMed]
27. Luo, T.; Xu, M.; Huang, T.; Ren, X.; Bu, X. Rethinking the intensified disparity in urbanization trajectory of a Chinese coastal province and its implications. *J. Clean. Prod.* **2018**, *195*, 1523–1532. [CrossRef]
28. Luo, X.; Ao, X.; Zhang, Z.; Wan, Q.; Liu, X. Spatiotemporal variations of cultivated land use efficiency in the Yangtze River Economic Belt based on carbon emission constraints. *J. Geogr. Sci.* **2020**, *30*, 535–552. [CrossRef]
29. Zhu, W.; Wang, M.; Zhang, B. The effects of urbanization on PM2.5 concentrations in China's Yangtze River Economic Belt: New evidence from spatial econometric analysis. *J. Clean. Prod.* **2019**, *239*, 118065. [CrossRef]
30. Zhang, Q.; Xu, C.Y.; Zhang, Z.; Chen, Y.D.; Liu, C.L. Spatial and temporal variability of precipitation over China, 1951–2005. *Theor. Appl. Climatol.* **2008**, *95*, 53–68. [CrossRef]
31. Sang, Y.-F.; Wang, Z.; Liu, C. Spatial and temporal variability of daily temperature during 1961–2010 in the Yangtze River Basin, China. *Quat. Int.* **2013**, *304*, 33–42. [CrossRef]
32. Karl, T.R.; Nicholls, N.; Ghazi, A. CLIVAR/GCOS/WMO Workshop on Indices and Indicators for Climate Extremes—Workshop summary. *Clim. Change* **1999**, *42*, 3–7. [CrossRef]
33. National Catalogue Service for Geographical Information. 1: 1 Million Public Version of Basic Geographic Information Data (2021). Available online: https://www.webmap.cn/commres.do?method=result100W (accessed on 12 May 2022).
34. Chen, Z.; Yu, B.; Yang, C.; Zhou, Y.; Yao, S.; Qian, X.; Wang, C.; Wu, B.; Wu, J. An extended time-series (2000–2018) of global NPP-VIIRS-like nighttime light data. *Earth Syst. Sci. Data* **2020**, *13*, 889–906. [CrossRef]
35. Ministry of Housing and Urban-Rural Construction of the People's Republic of China. China Urban Construction Statistical Yearbook. Available online: https://navi.cnki.net/knavi/yearbooks/YCJTJ/detail?uniplatform=NZKPT (accessed on 12 May 2022).
36. Zhang, L.; Ren, G.-Y.; Ren, Y.-Y.; Zhang, A.-Y.; Chu, Z.-Y.; Zhou, Y.-Q. Effect of data homogenization on estimate of temperature trend: A case of Huairou station in Beijing Municipality. *Theor. Appl. Climatol.* **2014**, *115*, 365–373. [CrossRef]
37. Menne, M.J.; Williams, C.N., Jr.; Palecki, M.A. On the reliability of the US surface temperature record. *J. Geophys. Res. Atmos.* **2010**, *115*, 11108. [CrossRef]

38. Trusilova, K.; Jung, M.; Churkina, G.; Karstens, U.; Heimann, M.; Claussen, M. Urbanization impacts on the climate in Europe: Numerical experiments by the PSU–NCAR Mesoscale Model (MM5). *J. Appl. Meteorol. Climatol.* **2008**, *47*, 1442–1455. [CrossRef]
39. Feng, J.-M.; Wang, Y.-L.; Ma, Z.-G. Long-term simulation of large-scale urbanization effect on the East Asian monsoon. *Clim. Change* **2015**, *129*, 511–523. [CrossRef]
40. Liao, W.; Wang, D.; Liu, X.; Wang, G.; Zhang, J. Estimated influence of urbanization on surface warming in Eastern China using time-varying land use data. *Int. J. Climatol.* **2017**, *37*, 3197–3208. [CrossRef]
41. Zhou, Y.; Smith, S.J.; Elvidge, C.D.; Zhao, K.; Thomson, A.; Imhoff, M. A cluster-based method to map urban area from DMSP/OLS nightlights. *Remote Sens. Environ.* **2014**, *147*, 173–185. [CrossRef]
42. Jiang, S.; Wang, K.; Mao, Y. Rapid Local Urbanization around Most Meteorological Stations Explains the Observed Daily Asymmetric Warming Rates across China from 1985 to 2017. *J. Clim.* **2020**, *33*, 9045–9061. [CrossRef]
43. Peterson, T.C.; Gallo, K.P.; Lawrimore, J.; Owen, T.W.; Huang, A.; McKittrick, D.A. Global rural temperature trends. *Geophys. Res. Lett.* **1999**, *26*, 329–332. [CrossRef]
44. Ye, H.; Huang, Z.; Huang, L.; Lin, L.; Luo, M. Effects of urbanization on increasing heat risks in South China. *Int. J. Climatol.* **2018**, *38*, 5551–5562. [CrossRef]
45. Shi, G.; Ye, P. Assessment on Temporal and Spatial Variation Analysis of Extreme Temperature Indices: A Case Study of the Yangtze River Basin. *Int. J. Environ. Res. Public Health* **2021**, *18*, 10936. [CrossRef] [PubMed]
46. Das, S.K.; Bebortta, S. A study on geospatially assessing the impact of COVID-19 in Maharashtra, India. *Egypt. J. Remote Sens. Space Sci.* **2022**, *25*, 221–232. [CrossRef]
47. Tomar, P.; Mishra, R.; Sheoran, K. Prediction of quality using ANN based on Teaching-Learning Optimization in component-based software systems. *Softw. Pract. Exp.* **2018**, *48*, 896–910. [CrossRef]
48. Krivoguz, D. Methodology of physiography zoning using machine learning: A case study of the Black Sea. *Russ. J. Earth Sci.* **2020**, *20*, 1–10. [CrossRef]
49. Ackom, E.K.; Adjei, K.A.; Odai, S.N. Spatio-temporal rainfall trend and homogeneity analysis in flood prone area: Case study of Odaw river basin-Ghana. *SN Appl. Sci.* **2020**, *2*, 1–26. [CrossRef]
50. Kumar, N.; Panchal, C.; Chandrawanshi, S.; Thanki, J. Analysis of rainfall by using Mann-Kendall trend, Sen's slope and variability at five districts of south Gujarat, India. *Mausam* **2017**, *68*, 205–222. [CrossRef]
51. Zhong, K.; Zheng, F.; Wu, H.; Qin, C.; Xu, X. Dynamic changes in temperature extremes and their association with atmospheric circulation patterns in the Songhua River Basin, China. *Atmos. Res.* **2017**, *190*, 77–88. [CrossRef]
52. Qiu, J.; Yang, X.; Cao, B.; Chen, Z.; Li, Y. Effects of Urbanization on Regional Extreme-Temperature Changes in China, 1960–2016. *Sustainability* **2020**, *12*, 6560. [CrossRef]
53. Yuan, Z.; Yin, J.; Wei, M.; Yuan, Y. Spatio-Temporal Variations in the Temperature and Precipitation Extremes in Yangtze River Basin, China during 1961–2020. *Atmosphere* **2021**, *12*, 1423. [CrossRef]
54. Guan, Y.; Zheng, F.; Zhang, X.; Wang, B. Trends and variability of daily precipitation and extremes during 1960–2012 in the Yangtze River Basin, China. *Int. J. Climatol.* **2017**, *37*, 1282–1298. [CrossRef]
55. Cheng, G.; Liu, Y.; Chen, Y.; Gao, W. Spatiotemporal variation and hotspots of climate change in the Yangtze River Watershed during 1958–2017. *J. Geogr. Sci.* **2022**, *32*, 141–155. [CrossRef]
56. Yu, R.; Xu, Y.; Zhou, T.; Li, J. Relation between rainfall duration and diurnal variation in the warm season precipitation over central eastern China. *Geophys. Res. Lett.* **2007**, *34*, 13703. [CrossRef]
57. Collier, C.G. The impact of urban areas on weather. *Q. J. R. Meteorol. Soc. J. Atmos. Sci. Appl. Meteorol. Phys. Oceanogr.* **2006**, *132*, 1–25. [CrossRef]
58. Johnson, B.; Shine, K.; Forster, P. The semi-direct aerosol effect: Impact of absorbing aerosols on marine stratocumulus. *Q. J. R. Meteorol. Soc.* **2004**, *130*, 1407–1422. [CrossRef]
59. Zhang, C.L.; Chen, F.; Miao, S.G.; Li, Q.C.; Xia, X.A.; Xuan, C.Y. Impacts of urban expansion and future green planting on summer precipitation in the Beijing metropolitan area. *J. Geophys. Res. Atmos.* **2009**, *114*, 02116. [CrossRef]

land

MDPI

Article

Influence of Urbanization on Spatio-Temporal Characteristics of Extreme Hourly Precipitation in Shenyang

Xue Ao [1], Qingfei Zhai [2], Chunyu Zhao [1,*], Yan Cui [1], Xiaoyu Zhou [1], Jingwei Li [1] and Mingqian Li [1]

1 Shenyang Regional Climate Center of Liaoning, Shenyang 110166, China
2 Liaoning Provincial Meteorological Information Center, Shenyang 110166, China
* Correspondence: springrainscn@163.com

Abstract: Understanding changes in extreme hourly precipitation is critical to urban planners for building more sustainable and resilient cities. In this study, we use satellite nighttime light data, urban land area data, population, and economic data to objectively classify urban and rural stations. Based on the hourly precipitation data from national meteorological stations in 1974–2020 and from regional automatic weather stations in 2005–2020 in Shenyang (China), the spatio-temporal distribution characteristics of the thresholds, maximums, intensities, and frequencies of extreme hourly precipitation (ExHP) in urban and rural areas are analyzed and compared. The results show that the large-value centers of ExHP thresholds, maximums, and intensities are mainly concentrated in urban areas. Both the frequency and intensity of ExHP at urban stations are obviously larger than those at rural stations, and the peaks mainly appear at night for stations of both two types. From 1974 to 2020, the average frequency and intensity of ExHP at urban stations both show increasing trends, with the increasing rate being much higher than those at rural stations. In terms of temporal variation, precipitation events of the abrupt type are the most frequent, accounting for 48.6% of the total, followed by the growing type (42.7%) and continuous type (8.7%). ExHP events of the abrupt type are mostly concentrated in Kangping County and Faku County of Northern Shenyang, but rarely occur in Xinmin City. ExHP events of the growing type are mainly found in Xinmin City and the municipal district of Shenyang. For urban stations, the ExHP frequency decreases in the early stage of urbanization, while increasing evidently during the rapid urbanization stage. However, the situation is just the opposite for rural stations. This indicates that the rapid urbanization in Shenyang has a certain impact on the increase in ExHP.

Keywords: urbanization; extreme hourly precipitation; Shenyang

1. Introduction

Under the background of global warming, the meteorological disasters in Liaoning Province (China) have been increasingly frequent, which brings a serious challenge to disaster prevention and mitigation [1,2]. From 2018 to 2019, 14 regional rainstorm processes occurred in Liaoning Province, with the emergency response to major meteorological disasters (rainstorms) being activated each time. A total of 14 cities were affected by the rainstorms, and the rainfall amount and rainfall days in many parts had set record highs, causing serious damages to transportation, animal husbandry, agriculture, and water conservancy projects. During 2018–2019, the number of typhoons affecting Liaoning Province is three times as many as that in the same period in history. These typhoons with strong intensity induced precipitation events with large rainfall amounts, which caused serious losses and continuously broke the extreme value of typhoon rainstorms [3]. On 16 August 2019 (local standard time, LST), the strongest short-term heavy rainfall in 68 years hit the municipal district of Shenyang with its intensity reaching the level of a heavy rainstorm, causing one death. From 1755 LST to 1855 LST, the precipitation observed at the meteorological station of Shenshuiwan Park in the Heping District reached

100.8 mm, which is the strongest hourly rainfall in the municipal district of Shenyang since complete meteorological observation records began in 1951. In August 2019, under the combined effect of Typhoon Lekima and the cold air brought by the upper-level trough, the regional-average precipitation amount in Liaoning Province reached 126.5 mm, breaking the record in the same period since 1951. This typhoon-induced precipitation event has been evaluated as a level 1 rainstorm disaster, which is the most serious for its long duration, large, accumulated rainfall amount, wide impact range, and accompanying strong gales. The rainstorm led to urban waterlogging in Shenyang, with the total area affected reaching 39.8 km^2 and a direct economic loss of 7.965 million yuan (1.669 million dollars) [4]. In addition, water conservancy projects and non-engineering measures for mountain torrents were also damaged to varying degrees. In recent decades, the impacts of rainstorm-induced mountain torrents, geological disasters, and urban waterlogging on Shenyang have become increasingly prominent. From 1984 to 2019, Shenyang suffered from varying degrees of meteorological disasters (including torrential rain, floods, and typhoons) year by year, resulting in a direct economic loss of more than 100 million yuan (14.65 million dollars) per year [5]. With the increasing population, social and economic development, and the continuous increases in carrying capacity and infrastructures per unit area in Shenyang, the damages caused by rainstorms are becoming more and more serious, resulting in great economic losses and a series of social and environmental problems. Hence, the rainstorm disaster has become a key factor restricting the sustainable development of Shenyang. Therefore, it is necessary to study the impact of urbanization on extreme heavy precipitation in this region.

According to the Clausius-Clapeyron equation, the water holding capacity of the atmosphere increases with the rising of atmospheric temperature. Therefore, the precipitation intensity will increase in the mid-latitudes and tropics which are both densely populated [6]. However, the precipitation intensity is usually expressed by daily precipitation data in previous studies, which may overestimate the intensity of long-term persistent weak precipitation and underestimate the intensity of short-term heavy precipitation, making the two types of precipitation events hard to be distinguished. Therefore, improving the temporal resolution of precipitation data is essential to studying extreme precipitation. In recent years, with the development of meteorological operation, a set of high-quality and relatively complete long-term hourly precipitation data has been established by Liaoning Meteorological Information Center, which is based on the observations at the national ground observation stations (61 stations in total, covering the national reference climatological stations, principal meteorological stations, and ordinary climatological stations) since records began and at the regional automatic stations (237 regional backbone stations in total) from 2005 to 2020 in Shenyang. This dataset provides important support for the study of extreme precipitation in Shenyang. So far, there have been some studies focusing on hourly precipitation in China. However, the research on extreme hourly precipitation (ExHP) events and urbanization effects is rarely seen. Hence, this dataset is essential to reveal the multi-scale spatio-temporal variation characteristics of ExHP in Shenyang.

Previous studies on extreme precipitation events and urbanization response are mainly concentrated in developed countries and highly developed regions in China, such as the Yangtze River Delta and Beijing-Tianjin-Hebei region. However, there are few studies on the urbanization effect of Shenyang, especially the impact of urbanization on extreme precipitation in this region. Studies have pointed out that the increase in urban buildings is conducive to convective precipitation and the enhancement of the urban heat island effect [7,8]. There is a significant relationship between the decrease in summer precipitation and the rapid urbanization in the Beijing-Tianjin-Hebei region [9–11]. Yang et al. [12,13] and Zhang et al. [14] suggested that the increasing trend of short-term heavy precipitation is closely related to urbanization. Moreover, the short-term heavy precipitation mostly occurs in the early morning, which may be caused by the joint action of the sea-land breeze and urbanization. Based on the TRMM (Tropical Rainfall Measuring Mission) satellite data, Li et al. [15] found that extreme heavy precipitation in the Pearl River Delta evidently increases compared with that in other regions, while weak precipitation significantly reduces. Dai et al. [16] revealed that the summer precipitation intensity and the hours of extremely heavy precipitation in Liaoning Province both show an increasing trend, where the intensity of extreme precipitation in July greatly increases. According to multi-model simulation results, Miao et al. [17] found that the precipitation distribution in urban areas is closely related to the degree of urbanization. Liang and Ding [18] analyzed the 100-year hourly precipitation data in Shanghai and found that the hourly extreme precipitation in Shanghai has an obvious increasing trend. Especially, in the 30 years of rapid urbanization since the 1980s, the rain island effect is obvious in Shanghai, where the increase in the magnitude of extreme precipitation is more obvious in the city center than in the suburbs. In the research on the characteristics of summer precipitation in Beijing and the impact of urbanization on heavy precipitation, Yuan et al. [19] suggested that the contribution of extremely heavy precipitation in urban areas to the total summer precipitation has been increasing since 2004, which may be due to the urbanization effect. However, the previous works on urban precipitation are not specific enough. Most of the research is based on a few national meteorological stations, and the analysis on an hourly scale is also relatively few. Fewer studies may be caused by the lack of density and high-quality observational data in urban areas, which is a necessity for investigating the relations between urbanization and precipitation events at a finer spatial and temporal scale.

In this study, multi-source data including nighttime light data, socio-economic data, and urban area are used to determine the thresholds to distinguish between urban and rural areas, so as to classify the stations into urban and rural stations. The hourly precipitation data at national meteorological stations from 1974 to 2020 and regional automatic weather stations from 2005 to 2020 in Shenyang are also used. The percentile method is adopted to determine the threshold of ExHP at each station. On this basis, the spatio-temporal distribution characteristics of threshold, maximum, intensity, and frequency of ExHP at urban and rural stations in Shenyang are analyzed and compared, aiming to provide technical support for extreme precipitation prediction, meteorological disaster prevention and mitigation, and the adaptation to urban climate change in Shenyang.

The remainder of this paper is organized as follows. Section 2 introduces the data and method. Section 3 illustrates the spatial distribution characteristics of ExHP in Shenyang. Section 4 presents the time variation characteristics of ExHP, such as diurnal variation, interannual variation, and the variation characteristics of ExHP events of different types. Finally, the conclusions and discussion are provided in Section 5.

2. Data and Methods

2.1. Study Area

Shenyang is a high-density mega-city covering about 1.3 million square kilometers, which is located in the central part of Liaoning Province of Northeast China. It is dominated by plains and has a temperate monsoon climate with annual average temperature and precipitation of 8.4 °C and 690 mm, respectively. The rainfalls in Shenyang are mainly concentrated in summer. Shenyang has a larger temperature difference and obvious seasonal characteristics including hot-humid summer and cold-dry winter. Shenyang includes Kangping County, Faku County, Xinmin City, Liaozhong District, and the municipal district of Shenyang. Because of urban expansion, the population and economy in Shenyang grow rapidly, over half of the population lives in the municipal district of Shenyang. Subregions of Shenyang present different urbanization patterns with the varied densities of city population and buildings as well.

2.2. Research Data

This study uses five types of research data. First, the hourly temperature and precipitation data at 7 national meteorological stations during 1974–2020 and 237 regional automatic stations since records began until 2020 in Shenyang are adopted (Figure 1), which are provided by the Liaoning Meteorological Information Center. The national meteorological stations are set up according to the national climate divisions to obtain long-term continuous climate data with sufficient representation. The number of stations is relatively small, while long-term observations. The regional automatic weather stations were established recently, but they provide more accurate and higher quality observations. They are thus an effective supplement to the national station data. Quality control procedures such as climatic extreme value and consistency tests have been applied to fix artificial errors. Second, the nighttime light data from the Defense Meteorological Satellite Program/Operational Line-Scan System (DMSP/OLS) with the study period during 1993–2013 are also utilized, which are provided by the National Geographic Data Center, https://www.ngdc.noaa.gov/eog/dmsp.html (accessed on 28 August 2022). The data is processed by accidental noise filtering and cloud elimination. Besides, it has also been specially developed with respect to the geographical characteristics of Asia, where the interference of light saturation is eliminated, and there is no necessity for manual control of the amplification gain. This data facilitates the analysis of urbanization intensity in Asia [20]. Third, the data on urban land area comes from the Liaoning Statistical Yearbook from 1993 to 2013. Fourth, the gross domestic product (GDP) per unit area adopts the 1-km grid GDP in China from the Resources and Environment Science and Data Center, http://www.resdc.cn (accessed on 28 August 2022). Fifth, the population density data comes from the Population Grid_China released by the Institute of Geographical Sciences and Natural Resources Research, Chinese Academy of Sciences. http://www.geodata.cn (accessed on 28 August 2022) . This dataset is established based on the land-use data from remote sensing data and population statistics. Through the spatial analysis function of the geographic information system, a spatial distribution model of the population is constructed to spatialize the statistical population data. Finally, the spatial population density data with a resolution of 1 km × 1 km is generated.

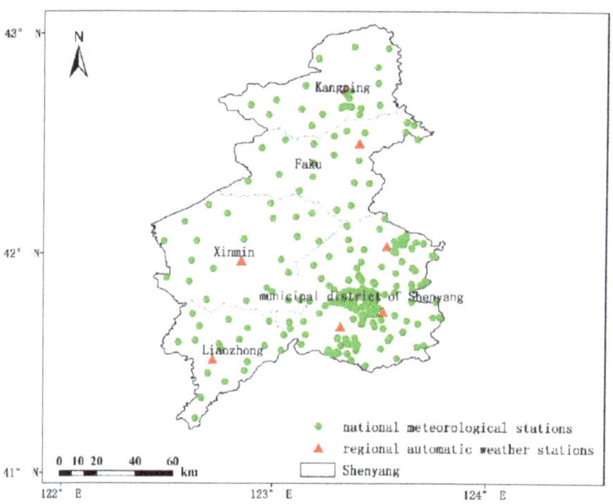

Figure 1. The spatial distribution of national meteorological stations and regional automatic weather stations in Shenyang.

2.3. Identification and Classification of Extreme Hourly Precipitation Events

In this study, the percentile method is used to determine the threshold of ExHP. The hourly precipitation data during 1974–2020 at each station are sorted from the weakest to the strongest. Through a comparison test, the hourly precipitation in the 99th percentile exceeding the minimum threshold of hourly heavy precipitation defined by the National Meteorological Center (20 mm$\cdot$h^{-1}) is defined as the threshold of ExHP at this station [21]. The frequency of ExHP is defined as the occurrence number of hourly precipitation exceeding the above-defined threshold. The intensity of ExHP is defined as the ratio of the total amount of ExHP at each station to the frequency of the ExHP.

Based on the relationships of the hourly precipitation three hours before the occurrence of ExHP (R_{-1}, R_{-2}, R_{-3}) with the ExHP (R_0), the ExHP events from 1974 to 2020 are classified into the abrupt type, growing type, and continuous type [22]. ExHP events of the abrupt type generally last for a short time, with a small range and strong intensity. However, ExHP events of the continuous type are generally featured by longer duration and larger precipitation range with little variation in the intensity, mostly from the nimbostratus and altostratus and often related to the frontal system.

Specifically, when the magnitudes of R_{-1}, R_{-2}, and R_{-3} are all less than one-tenth of that of R_0 (that is, $R_{-1} < 0.1 \times R_0$, $R_{-2} < 0.1 \times R_0$, and $R_{-3} < 0.1 \times R_0$), the precipitation event is defined as the abrupt type. When $R_{-1} > R_{-2}$ or $R_{-1} > R_{-3}$ or $R_{-2} > R_{-3}$ and at least one of (R_{-1}, R_{-2}, R_{-3}) has a magnitude greater than one-tenth of that of R_0 but smaller than R_0, it is defined as a growing type. When at least one of (R_{-1}, R_{-2}, R_{-3}) has a magnitude greater than that of R_0, it is defined as a continuous type [22].

The linear trend estimation is adopted for climate change analysis. The regression coefficient b indicates the trend of climate variable x, where $b > 0$ indicates that x exhibits an upward trend with the increase in time t and $b < 0$ stands for the downward trend of x with the increase in time t. The magnitude of b reflects the changing rate, namely the degree of the upward or downward tendency [23].

In this study, the non-parametric statistical method is used for the significance test of linear trend [24], and the statistical formula is as follows:

$$Z = \frac{4}{n(n-1)} \sum_{i=1}^{n-1} r_i - 1 \tag{1}$$

$$Z_{0.05} = 1.96 \left[\frac{4n+10}{9n(n-1)} \right]^{\frac{1}{2}} \tag{2}$$

where r_i in Equation (1) is the climatic sequence. If $|Z| > Z_{0.05}$, it is considered that the change trend is significant at the significance level of $\alpha = 0.05$.

2.4. Division of Urban and Rural Stations

By comparing with the statistical data, the spatial information of urban land use in Liaoning Province is extracted based on the DMSP/OLS nighttime light data [25–27]. The extraction mainly follows the two assumptions below. First, it is considered that the data on the urban land area in the statistical yearbook can accurately reflect the area of urban land in Shenyang, so the urban land area extracted from the DMSP/OLS data should be as close as possible to the data in the statistical yearbook. Second, the urban land area in Shenyang has expanded continuously since the 1990s. It is considered that the grid representing the urban area in the previous period does not disappear in the next period, that is, the grid in the previous period should be retained in the next period. On this basis, the dynamic thresholds are determined through the method of bisection, and the urban area under different thresholds is calculated. If the urban area calculated by the nighttime light data under a certain threshold is the closest to that in the statistical yearbook, this threshold is determined as the optimal threshold in Shenyang. After that, the average gray value of nighttime light within a radius of 7 km from each station is calculated. If the average value is greater than the threshold of the city where the station is located, this station is regarded as an urban station [28,29].

Figure 2 is a flowchart of the urban station and rural station divisions. D(i) is the threshold value of DMSP/OLS nighttime light data in a certain year (1 < D(i) < 63). Its initial value is set as 1 while i is the number of iterations. S(i) is the urban area of nighttime light of the city in this year when the threshold is equal to D(i), (S(i) is equal to the number of pixels × pixel area in an urban area). Area indicates the statistical value of the area of the city in the corresponding year, while P(i) is the average gray value of nighttime light within a radius of 7 km from each station.

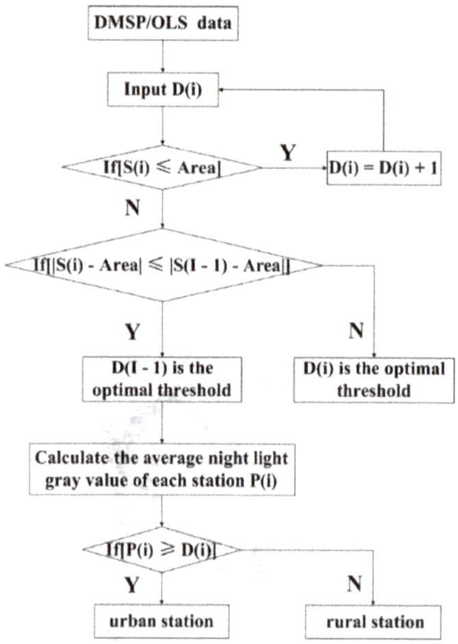

Figure 2. A flowchart of urban station and rural station division.

The spatial distribution of nighttime light gray values (Figure 3) shows that the gray values of some areas in the municipal district of Shenyang and Liaozhong District have increased from 1993 to 2003, indicating that these areas have undergone the urbanization process significantly. Combined with the population density data and the GDP per unit area (Figure 4), it can be seen that the urbanization process in Kangping County, Faku County, and Xinmin City is relatively slower. Therefore, the national meteorological stations and regional automatic weather stations in Liaozhong District and in the municipal district of Shenyang are selected as urban stations, and the rest of the stations are rural.

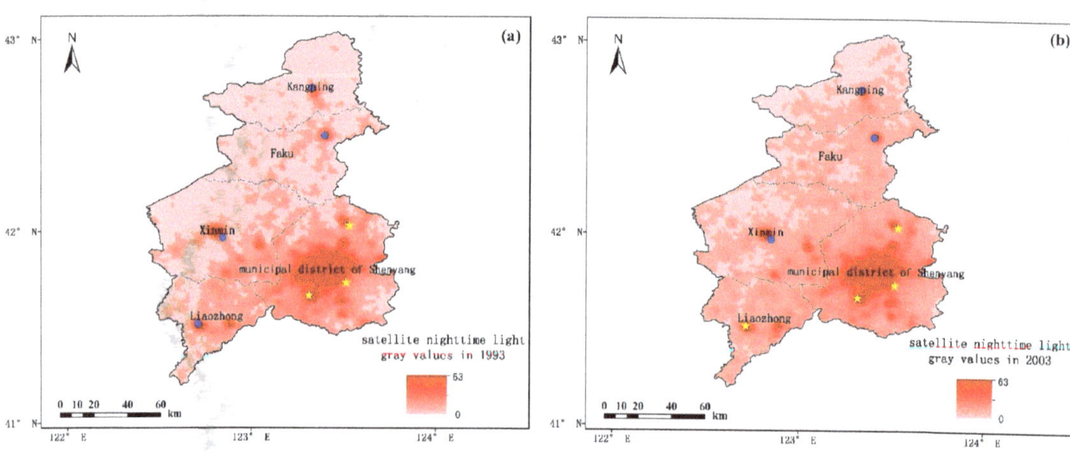

Figure 3. *Cont.*

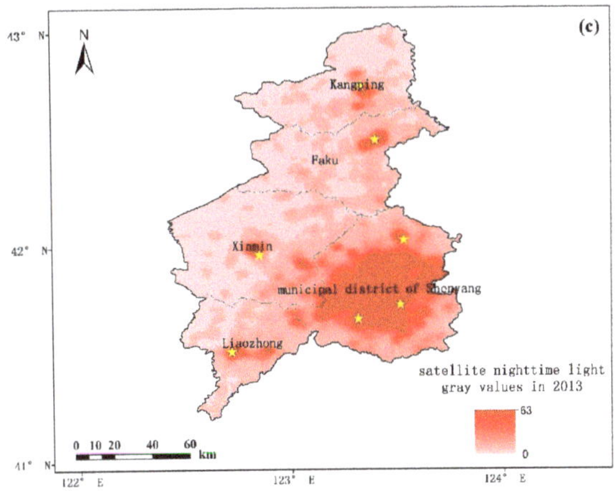

Figure 3. Spatial distributions of satellite nighttime light gray values in Shenyang in (**a**) 1993, (**b**) 2003, and (**c**) 2013 (yellow dots represent urban stations and blue dots represent rural stations).

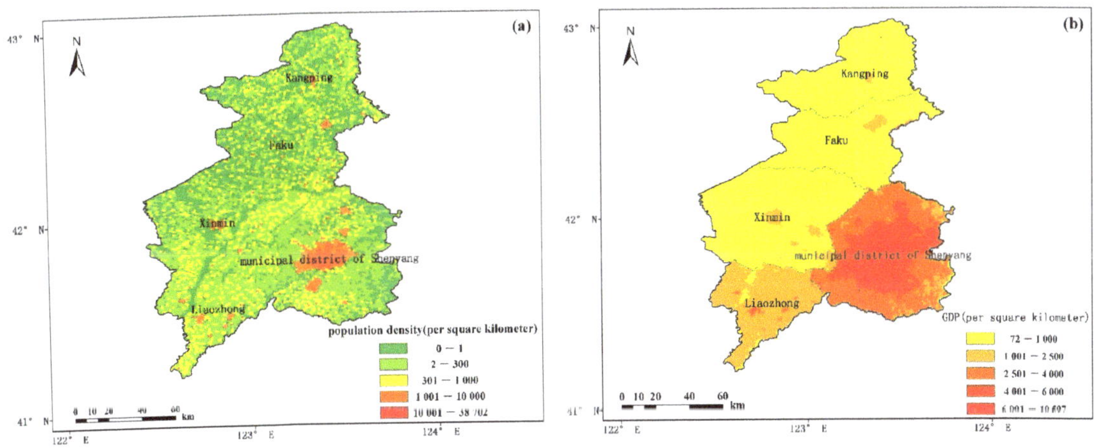

Figure 4. Spatial distributions of (**a**) population density and (**b**) gross domestic product per unit area in Shenyang.

2.5. Definition of Urbanization Impact

To quantitatively evaluate the urbanization impact on precipitation, the following terms are defined according to the Chu and Ren [30].

The linear trend of meteorological elements at stations near the city caused by urbanization factors is called the urbanization impact (ΔX_{ur}), which is represented by the difference in the change trend of meteorological elements between urban stations (X_u) and rural stations (X_r):

$$\Delta X_{ur} = X_u - X_r \tag{3}$$

The contribution rate of urbanization impact refers to the proportion of urbanization impact to the change trend of meteorological elements at urban stations, which is expressed by E_u:

$$E_u = |\Delta X_{ur}/X_u| \times 100\% = |(X_u - X_r)/X_u| \times 100\% \tag{4}$$

3. Results

3.1. Spatial Distribution Characteristics of Extreme Hourly Precipitation in Shenyang

The spatial distributions of ExHP at national and regional stations (Figure 5) show that the large-value centers of the thresholds, maxima, and intensities of ExHP are mainly concentrated in the areas where urban stations are located. The maximum value of the ExHP threshold (32.43 mm·h^{-1}) is located at Beiling Park station in Huanggu District, followed by 29.9 mm·h^{-1} at Zhujia Street station in Hunnan District. Table 1 reveals that all the large-value areas of ExHP thresholds are concentrated in the municipal district of Shenyang. Especially among the stations with the top 10 ExHP thresholds, 5 stations are located in the Sujiatun District. The areas with large ExHP thresholds are mainly located in both the southern parts of the municipal district of Shenyang and Liaozhong District. The maximum ExHP (115.7 mm·h^{-1}) is found in Kangping County at 1900 LST (Local Standard Time) on 16 July 1988, followed by 111 mm·h^{-1} in Liaozhong District at 2100 LST on 3 July 1991, and 100.8 mm·h^{-1} in Shenshuiwan Park of Heping District at 1800 LST on 16 August 2019. Combined with Table 1, it is revealed that the large values of ExHP maximums are mainly located in the municipal district of Shenyang, which mostly occur at night. Among the stations with the top 10 ExHP maximums, five stations are located in the municipal district of Shenyang and two in Liaozhong District. The large values of ExHP maximums are mainly concentrated in the southeastern part of the municipal district of Shenyang, western Liaozhong District, and northern Kangping County, while the values of ExHP maximums in Faku County and Xinmin City are relatively smaller. The maximum value of ExHP intensity (43.1 mm·h^{-1}) is located at Yangguang station in Yuhong District, followed by 42.3 mm·h^{-1} at Beiling Park station of Huanggu District. Combined with Table 2, it can be seen that the large-value areas of ExHP intensity are all located in the municipal district of Shenyang. In addition, among the stations with the top 10 ExHP intensities, three stations are located in Huanggu District. Moreover, the spatial distribution also displays that the large values of ExHP intensity are mainly concentrated in the southeastern municipal district of Shenyang and the west part of Liaozhong District. In terms of spatial distribution, the ExHP frequency differs greatly from the threshold, intensity, and maximum. The ExHP frequency at urban stations is relatively low, whereas the large-value areas are mainly in Xinmin City. These may be due to the difference in construction time between national and regional stations. The hourly precipitation records at the national principal meteorological stations and reference climatological stations in Shenyang began in 1974, with few missing reports. However, the regional automatic weather stations have been gradually put into operation since 2003, which were established in different periods and have missing reports to varying degrees. Therefore, the top 10 ExHP frequencies are basically found at the national principal meteorological stations and reference climatological stations, and the maximum value (119 times) is located in Xinmin City.

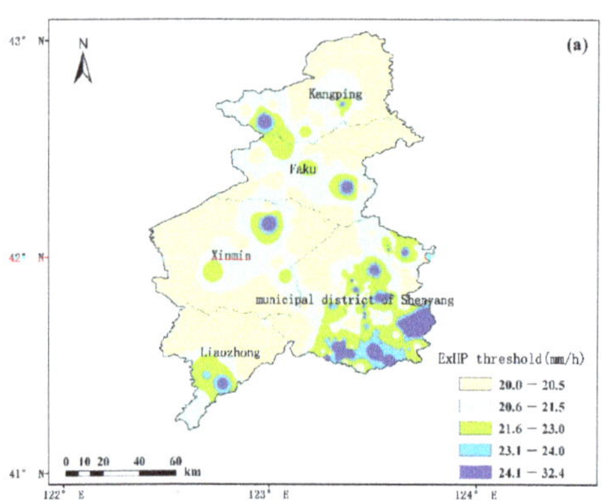

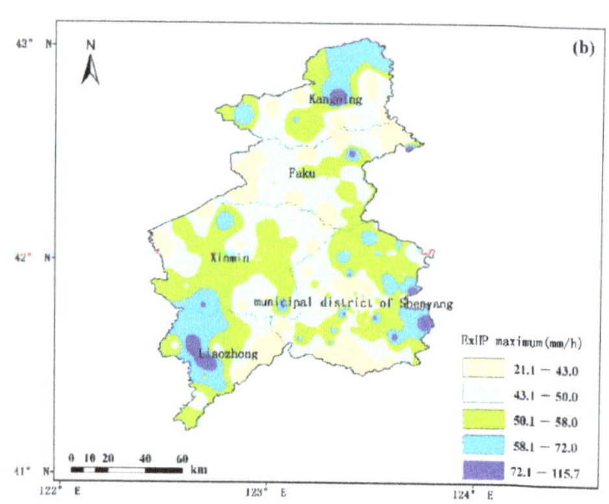

Figure 5. *Cont.*

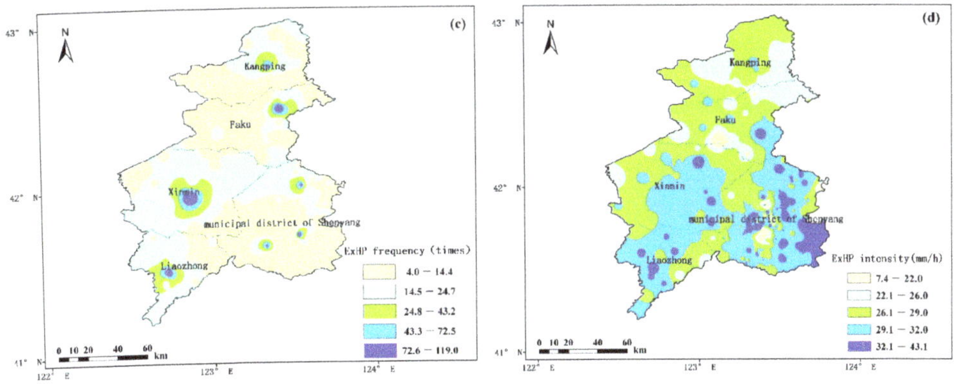

Figure 5. Spatial distributions of (**a**) extreme hourly precipitation (ExHP) threshold, (**b**) ExHP maximum, (**c**) ExHP frequency and (**d**) ExHP intensity.

Table 1. National and regional stations with the top 10 ExHP thresholds and maximums.

County (District)	Station Name	Station Number	Threshold (mm·h^{-1})	County (District)	Station	Station Number	Maximum (mm·h^{-1})
Huanggu District	Beiling Park	L1138	32.43	Kangping County	Kangping	54244	115.7
Hunnan District	Zhujia Street	L1089	29.901	Liaozhong District	Liaozhong	54332	111
Sujiatun District	Linsheng Street	L1103	29.79	Heping District	Shenshuiwan Park	L1310	100.8
Sujiatun District	Linsheng Coal Mine	L1174	29.359	Hunnan District	Gaokan Street	L1065	89.5
Huanggu District	Santaizi Overpass	L1274	28.288	Liaozhong District	Manduhu Town	L1100	86.7
Hunnan District	Wangbin Street	L1085	28.04	Faku County	Dangjiajie Village of Bojiagou Town	L1165	85
Sujiatun District	Changxingdian Village of Linsheng Street	L1175	28	Hunnan District	Wangbin Street	L1085	83.7
Xinmin City	Gongzhutun Town	L1039	26.7	Yuhong District	Yangguang	L1255	83.4
Sujiatun District	Yantai Village of Bayihongling Street	L1170	26.43	Hunnan District	Shenyang	54342	82.2
Sujiatun District	Chenxiang Street	L1109	26.375	Faku County	Faku	54245	79.8

Table 2. National and regional stations with the top 10 ExHP frequencies and intensities.

County (District)	Station	Station Number	Frequency (Times)	County (District)	Station	Station Number	Intensity (mm·h^{-1})
Xinmin City	Xinmin	54333	119	Yuhong District	Yangguang 100	L1255	43.1
Faku County	Faku	54245	118	Huanggu District	Beiling Park	L1138	42.3
Hunnan District	Shenyang	54342	117	Tiexi District	Huaxiang Primary School	L1259	41.9
Liaozhong District	Liaozhong District	54332	106	Shenhe District	Dongling Street	L1124	39.3
Shenbei New District	Shenbei New District	54248	91	Hunnan District	Wangbin Street	L1085	38.3
Sujiatun District	Sujiatun	54340	82	Shenfu Demonstration District	Shenjingzi Street	L1081	37.9
Kangping County	Kangping	54244	76	Tiexi District	Nanba Bridge	L1126	37.7
Liaozhong District	Yujiafang Town	L1120	24	Huanggu District	Beitabeng	L1261	37.6
Shenbei New District	Huishan Street	L1054	24	Huanggu District	Santaizi Overpass	L1274	37.6
Faku County	Yiniubaozi Town	L1034	24	Shenhe District	Kepu Park	L1275	37.3

3.2. Temporal Variation Characteristics of Extreme Hourly Precipitation

3.2.1. Diurnal Variation Characteristics

The diurnal variations of the regional-averaged ExHP frequency and intensity at urban and rural stations in Shenyang are shown in Figure 6. It is revealed that for urban stations, the diurnal variations of ExHP frequency and intensity both exhibit a bimodal distribution. The ExHP frequency peaks at 1800 LST and 2100 LST, while the ExHP intensity peaks in the afternoon (1300 LST) and at night (2100 LST). The ExHP frequency and intensity before dawn and in the early morning are significantly lower than those in the afternoon and at night. While for rural stations, the ExHP frequency and intensity mainly peak at night (2100 LST). It may be attributed to the enhancement of the southwest jet at night, which promotes water vapor transport, resulting in the frequent occurrence of short-term heavy rainfall at night [31]. To sum up, both the ExHP frequency and intensity at urban stations are much larger than those at rural stations, and both urban and rural stations mainly peak at night.

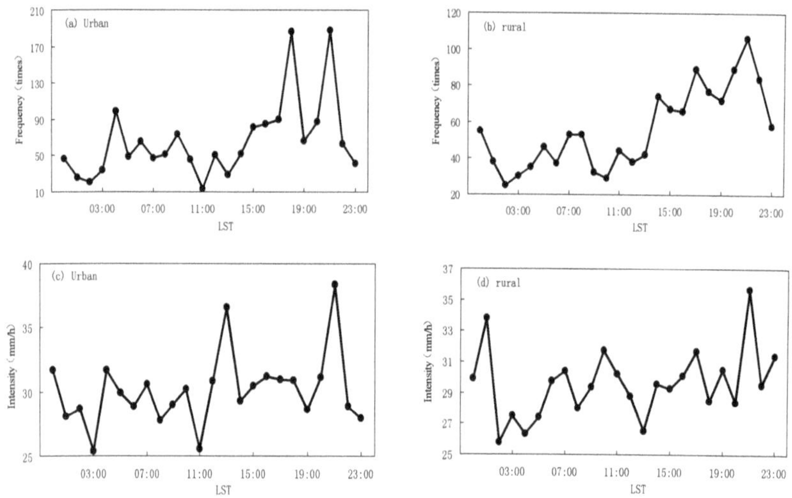

Figure 6. Diurnal variations of regional-averaged ExHP (**a**,**b**) frequency and (**c**,**d**) intensity at urban and rural stations in Shenyang.

3.2.2. Interannual Variation Characteristics

Figure 7 displays that both the ExHP frequency and intensity averaged at urban stations in Shenyang show increasing trends from 1974 to 2020, which are 0.17 times$\cdot$10a^{-1} and 0.76 mm$\cdot$h$^{-1}\cdot$10a^{-1}, respectively, where the variation trend of ExHP intensity has passed the significance test. However, there is no obvious variation trend of regional-averaged ExHP frequency at rural stations, and the ExHP intensity shows an upward trend with the trend value being 0.41 mm$\cdot$h$^{-1}\cdot$10a^{-1}. It can be seen that the increasing rate of ExHP intensity at urban stations is about twice that of rural stations. For both ExHP frequency and intensity, the increasing rates at urban stations are obviously larger than those at rural stations. The changes at rural stations are relatively gentle, indicating that rural stations can well represent the change of background field and the developing rate of areas not affected by urbanization, and thus the selection of rural stations is reasonable. Table 3 shows the urbanization effect on precipitation in Liaoning Province from 1974 to 2020. It can be found that urbanization has a great impact on precipitation intensity, and the contribution rate of urbanization reaches 46.1%, indicating that the increase in ExHP intensity in Liaoning Province in the last 47 years might be attributed to the impact of rapid urbanization.

Table 3. Urbanization impact and its contribution rate (%) on annual extreme hourly precipitation frequency and intensity at urban and rural stations in Liaoning Province from 1974 to 2020.

Type of Station	ExHP Frequency (Times$\cdot$(10a)$^{-1}$)	ExHP Intensity (mm$\cdot$h$^{-1}\cdot$(10a)$^{-1}$)
Urban station	0.17	0.76 *
Rural station	−0.004	0.41 *
Urbanization impact	0.174	0.35 *
Contribution rate of Urbanization impact (%)	-	46.1

Note: "*" means that the value has passed the significance test at 0.05 significance level; "-" means that the urbanization impact fails the significance test at 0.05 significance level, and then the contribution rate of urbanization impact will not be calculated.

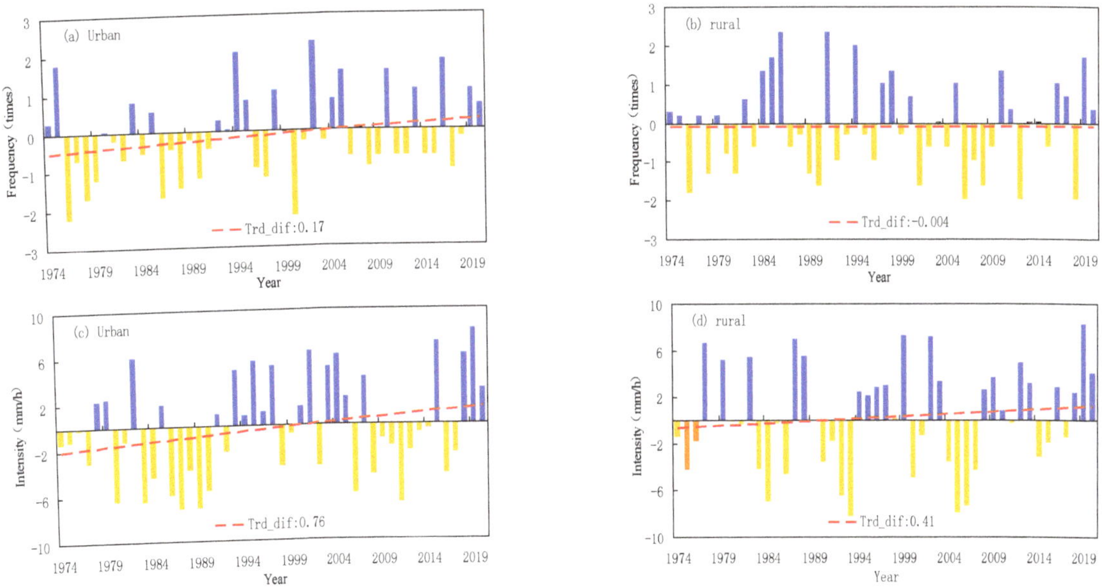

Figure 7. Interannual variations of regional-averaged ExHP (**a**,**b**) frequency and (**c**,**d**) intensity at urban and rural stations in Shenyang from 1974 to 2020.

To explore the variation of ExHP during the urbanization process in Shenyang, it is necessary to divide the development process in Shenyang into different periods. Economic development and population growth in urban regions are the main signs of urbanization. The statistical yearbooks show that the growth rate of the non-agricultural population in Shenyang is 2.7%·(10a)$^{-1}$ from 1974 to 1990. After the 1990s, the non-agricultural population increases continuously with the rapid development of the urban social economy in Shenyang. From 1991 to 2020, the growth rate of the non-agricultural population in Shenyang reaches 9.78%·(10a)$^{-1}$, and meanwhile, the GDP begins to grow rapidly from the 1990s. Therefore, the period before the early 1990s is selected as the early stage of urbanization in Shenyang in this study, while the period after the 1990s is the rapid development stage. The ExHP frequency at urban stations basically decreases in the early stage of urbanization (1974–1990), while increases evidently in the rapid development stage of urbanization (1991–2020) (Figure 8a–d). The ExHP frequency at rural stations increases in the early stage but decreases in the rapid development stage (Figure 8e–g). From 1974 to 2020, the change rates at the four urban representative stations are 0.47 times·(10a)$^{-1}$, 0.25 times·(10a)$^{-1}$, 0.55 times·(10a)$^{-1}$ and 0.41 times·(10a)$^{-1}$, respectively. Meanwhile, the change rates at the three rural representative stations are −0.02 times·(10a)$^{-1}$, 0.14 times·(10a)$^{-1}$, and −0.03 times·(10a)$^{-1}$, respectively. Meanwhile, a sensitivity experiment is carried out on the selected time nodes (1991 mentioned above) through a comparison between the previous and subsequent periods. It is found that the conclusions are basically the same no matter which year from 1990 to 1993 is selected as the time node. This indicates that the conclusions can truly reflect the differences in the variation characteristics of ExHP in different urbanization stages. In a word, the rapid urbanization in Shenyang has a certain impact on the increase in ExHP.

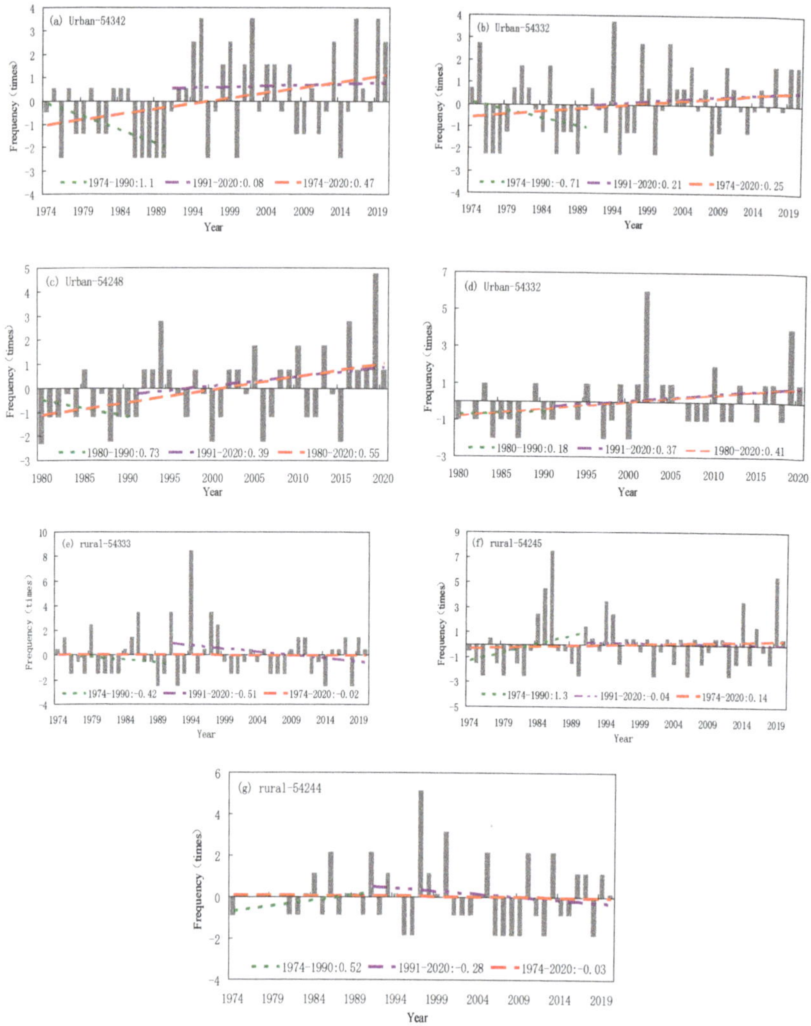

Figure 8. Annual ExHP frequencies at urban representative stations (**a**–**d**) and rural representative stations (**e**–**g**) in Shenyang from 1974 to 2020.

3.2.3. Variation Characteristics of Different Types of Extreme Hourly Precipitation Events

According to the evolution characteristics of precipitation three hours before the occurrence of ExHP, ExHP events are divided into the abrupt type, growing type, and continuous type (Table 4). Based on the frequencies of different types of ExHP events averaged in the whole of Shenyang City from 1974 to 2020, ExHP events of the abrupt type are the most, accounting for 48.6% of the total events, followed by the growing type (42.7%). There are relatively few ExHP events of the continuous type, accounting for only 8.7%.

Figure 9 shows the spatial distributions of the proportions of three types of ExHP events at national and regional stations. It is revealed that the abrupt type and growing type present opposite spatial distribution patterns. ExHP events of the abrupt type are prone to occur in Kangping and Faku Counties of northern Shenyang, but rarely in Xinmin City. However, ExHP events of the growing type occur more frequently in Xinmin City and the municipal district of Shenyang, but less in Kangping and Faku Counties. While ExHP events of the continuous type are mainly concentrated in northeastern Xinmin City and Hunnan District, but rarely occur in Kangping County, Faku County, Tiexi District, and Sujiatun District.

Table 4. Classification criteria and proportions of different types of ExHP events in Shenyang.

Type	Classification Criteria	Proportion
abrupt type	$R_{-1} < 0.1 \times R_0$, $R_{-2} < 0.1 \times R_0$ and $R_{-3} < 0.1 \times R_0$	48.6%
growing type	When $R_{-1} > R_{-2}$ or $R_{-1} > R_{-3}$ or $R_{-2} > R_{-3}$ and at least one of (R_{-1}, R_{-2}, R_{-3}) has a magnitude greater than one-tenth of that of R_0 but smaller than R_0	42.7%
continuous type	When at least one of (R_{-1}, R_{-2}, R_{-3}) has a magnitude greater than that of R_0	8.7%

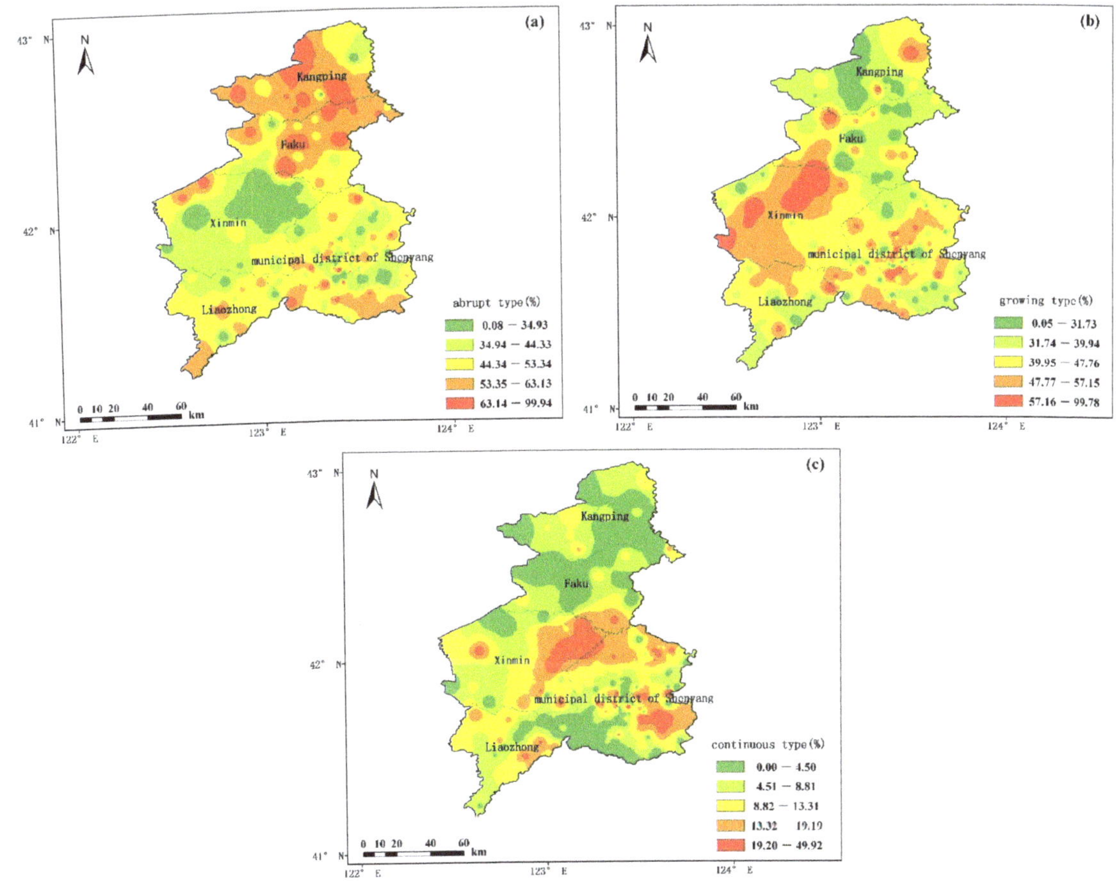

Figure 9. Spatial distributions for ExHP events of the (**a**) abrupt type, (**b**) growing type, and (**c**) continuous type in Shenyang.

4. Discussion

In this study, we evaluated the ExHP changes using the multi-source datasets by classifying the urban and rural stations, and in turn investigating the influences of urbanization on the ExHP. Results clearly indicate that the ExHP for urban and rural stations peaks in the nighttime, while urban stations have larger ExHP frequency and intensity than rural stations. It may be attributed to the enhancement of the southwest jet at night, which can promote water vapor transport and result in the frequent occurrence of short-term heavy rainfall at night [31]. Relatively, due to the diurnal variation of solar radiation heating, the low-level atmosphere is easy to be unstable during the afternoon and evening, benefiting local moist convection activities and short-term heavy precipitation. Meanwhile, the strong nighttime heat island effect produced by rapid urbanization, combined with the slowdown of the movement of the weather system by buildings and the increase of condensation nuclei, have enhanced nighttime precipitation [32–34]. Some studies believe that the nighttime precipitation peak comes from stratiform cloud precipitation, which is strengthened by the instability caused by the nighttime radiative cooling of the cloud top, the accumulation of water vapor at the low level greatly promotes the development of nighttime convection [35,36].

This study found that the large-value centers of ExHP thresholds, maximums, and intensities are mainly located in urban areas. Both the ExHP frequency and intensity averaged at urban stations in Shenyang show increasing trends from 1974 to 2020, which may be related to the urbanization of Shenyang. Li et al. [37] found that there are obvious high-value centers in the urban area and its vicinities, such as the hourly precipitation, precipitation intensity, and the frequency of short-term heavy precipitation in Beijing in summer, which is consistent with the results in this paper. Liu [38] also pointed out the main urban area of Beijing has an obvious "urban rain island effect" from July to August and the hourly precipitation and precipitation intensity are significantly greater than those in the suburbs. Yin et al. [39] pointed out that the increase in summer precipitation in the urban area of Beijing may be caused by the combined effect of the urban heat island circulation and the wind in the valley. Local sensible and latent heat transport in urban areas enhances geopotential instability in the lower atmosphere, directly triggering convection [40]. Surface temperature plays the most important role in enhancing sensible heat transport and triggering precipitation, and the heterogeneity of urban underlying surfaces enhances the locality of precipitation [41]. The water vapor of local convective precipitation comes from the evaporation of the surface and the upward transport of water vapor when the low-level wind field converges [42]. The impact of cities on climate is not limited to the above aspects. The urbanization effect changes the local circulation in the city and in turn affects the distribution and intensity of precipitation [43].

Since this study is focused on urbanization rates affect ExHP, it might be interesting to study how this is likely to evolve in the next decades. Some studies have used spatial and transient modelling of land use/land cover (LULC) to simulate and predict the changes in land use/land cover and population growth in the future. The model can predict spatial and temporal land changes for the next decades. It can also be a framework to understand underlying drivers of changes [44]. In the next step, we can use the model to assess the urbanization rates of Shenyang in the future, which can be used to predict the trend of ExHP and provide recommendations for urban planning.

A recent study using WAF model simulations shows that differences in the timing for urban-induced rainfall in the inland versus coastal environment, and the shape of a city/urban area can significantly affect and increase rainfall in this city. A circular city shows nearly 22.0% larger rainfall accumulation and 78.6% greater rainfall intensity compared to a triangular city over urban surfaces. The rainfall anomaly is caused by different urban-rural circulations over various city shapes. The strong low-level convergence over circular cities favors efficient upward moisture transport and deep convection [45]. The municipal district of Shenyang is a circular inland urban area, this study explains the reason for the increase in ExHP in the urban area. In future research work, we need to further explore the processes and mechanisms of the interaction between urbanization and precipitation through climate models and other technical methods. In order to provide technical support for extreme precipitation prediction, meteorological disaster prevention and mitigation, and the adaptation to urban climate change.

5. Conclusions

The large-value centers of ExHP thresholds, maximums, and intensities are mainly located in urban areas. The maximum ExHP threshold appears at Beiling Park station in Huanggu District with the value being 32.43 mm·h^{-1}. The maximum ExHP is 115.7 mm in Kangping County. The maximum ExHP intensity is 43.1 mm·h^{-1} at Yangguang station in Yuhong District. The top 10 ExHP frequencies mainly occur at national principal meteorological stations and reference climatological stations, and the maximum value (119 times) appears in Xinmin City. The diurnal variations of ExHP frequency and intensity at urban stations both exhibit a bimodal distribution. The ExHP frequency and intensity before dawn and in the early morning are significantly lower than those in the afternoon and at night. Both the ExHP frequency and intensity at urban stations are much larger than those at rural stations, and both urban and rural stations mainly peak at night. From 1974 to 2020, both the ExHP frequency and intensity averaged at urban stations in Shenyang show increasing trends, with the increasing rate being obviously larger at urban stations than at rural stations.

According to the statistical yearbooks, the period before the early 1990s is selected as the early stage of urbanization in Shenyang, and the period after the 1990s is the rapid development stage. Thus, the ExHP frequency at urban stations basically decreases in the early stage of urbanization and increases evidently in the rapid development stage of urbanization. While the ExHP frequency at rural stations increases in the early stage of urbanization but decreases in the rapid development stage. The contribution rate of urbanization on ExHP intensity reaches 46.1%, indicating that rapid urbanization has a certain impact on the increase in ExHP in Liaoning Province.

From 1974 to 2020, ExHP events of the abrupt type are the most, accounting for 48.6% of the total ExHP events, followed by the growing type (42.7%). While there are relatively few ExHP events of the continuous type, accounting for only 8.7%. ExHP events of the abrupt type are prone to occur in Kangping and Faku Counties of northern Shenyang, but rarely in Xinmin City. However, ExHP events of the growing type occur more frequently in Xinmin City and the municipal district of Shenyang. Besides, ExHP events of the continuous type are mainly concentrated in northeastern Xinmin City and Hunnan District.

Author Contributions: Conceptualization, X.A. and C.Z.; methodology, Q.Z.; formal analysis, Y.C.; data curation, X.Z.; writing—original draft preparation, X.A.; writing—review and editing, Q.Z.; visualization, J.L.; supervision, M.L. All authors have read and agreed to the published version of the manuscript.

Funding: This research was funded by the Scientific research project of liaoning Meteorological Bureau in 2021 (No. 202108) and the Science and Technology Plan Project in Liaoning Province (No. 2019-MS-199).

Institutional Review Board Statement: Not applicable.

Informed Consent Statement: Not applicable.

Data Availability Statement: Some models, or code that support the findings of this study are available from the corresponding author upon reasonable request.

Conflicts of Interest: The authors declare no conflict of interest.

References

1. Chen, Y.F.; Gao, G. An analysis to losses caused by meteorological disasters in China during 1989–2008, China. *Meteorol. Mon.* **2010**, *36*, 76–80.
2. Chen, N.N.; Zhang, Y.S.; Zhang, S.; Liu, J.; Wang, T.; Xu, Y.Q. Analysis of main meteorological disasters distribution and early warning service in Liaoning Province from 2004 to 2013, China. *J. Nat. Disaster* **2017**, *26*, 176–184. [CrossRef]
3. China Meteorological Administration. *Yearbook of Meteorological Disaster in China*; China Meteorological Press: Beijing, China, 2020.
4. China Meteorological Administration. *Yearbook of Meteorological Disaster in China*; China Meteorological Press: Beijing, China, 2021.
5. Li, B.; Meng, Q.N. *Meteorological Disaster Ceremony of China-Liaoning Volume*; Meteorological Press: Nanjing, China, 2005.
6. IPCC. *Climate Change 2013: The Physical Science Basis*; Cambridge University Press: Cambridge, UK, 2013.

7. Yang, P.; Ren, G.Y.; Yan, P.C.; Deng, J.M. Urbanization signifcantly reduces frequency of light rain: An example from Beijing City. *Teor. Appl. Climatol.* **2021**, *145*, 763–774. [CrossRef]
8. Feng, J.M.; Wang, Y.L.; Ma, Z.G.; Liu, Y.H. Simulating the regional impacts of urbanization and anthropogenic heat release on climate across China. *J. Clim.* **2012**, *25*, 7187–7203. [CrossRef]
9. Lu, R. Linear relationship between the interdecadal and interannual variabilities of North China rainfall in rainy season. *Chin. Sci. Bull.* **2003**, *48*, 1040–1044. [CrossRef]
10. Wei, F.Y.; Yu, X.; Mann, M.E. Probabilistic trend of anomalous summer rainfall in Beijing: Role of interdecadal variability. *J. Geophys. Res.* **2008**, *113*, D20106. [CrossRef]
11. Zhang, C.L.; Chen, F.; Miao, S.G.; Li, Q.C.; Xia, X.A.; Xuan, C.Y. Impacts of urban expansion and future green planting on summer precipitation in the Beijing metropolitan area. *J. Geophys. Res.* **2009**, *114*, D02116. [CrossRef]
12. Yang, P.; Ren, G.Y.; Hou, W.; Liu, W.D. Spatial and diurnal characteristics of summer rainfall over Beijing municipality based on a high-density AWS dataset. *Int. J. Climatol.* **2013**, *33*, 2769–2780. [CrossRef]
13. Yang, P.; Ren, G.Y.; Yan, P.C. Evidence for a strong association of short-duration intense rainfall with urbanization in the Beijing urban area. *J. Clim.* **2017**, *30*, 5851–5870. [CrossRef]
14. Zhang, Y.; Smith, J.A.; Luo, L.F., Wang, Z.F.; Baeck, M.L. Urbanization and rainfall variability in the Beijing metropolitan region. *J. Hydrometeorol.* **2014**, *15*, 2219–2235. [CrossRef]
15. Li, W.B.; Chen, S.; Chen, G.X.; Sha, W.M.; Wang, B.-M. Urbanization signatures in strong versus weak precipitation over the Pearl River Delta metropolitan regions of China. *Environ. Res. Lett.* **2011**, *6*, 049503. [CrossRef]
16. Dai, T.R.; Wang, X.T.; Luan, J.; Wang, S. Analysis on Spatial and temporal distribution characteristics of precipitation in summer in Liaoning Province during 1961-2013, China. *Mod. Agric. Sci. Technol.* **2016**, *10*, 205–209.
17. Miao, S.; Chen, F.; Li, Q.C.; Fan, S.Y. Impacts of urban processes and urbanization on summer precipitation: A case study of heavy rainfall in Beijing on 1 August 2006. *J. Appl. Meteorol. Clim.* **2011**, *50*, 806–825. [CrossRef]
18. Liang, P.; Ding, Y.H.; He, J.H.; Tang, X. Study of relationship between urbanization speed and change in spatial distribution of rainfall over Shanghai. *J. Trop. Meteorol.* **2013**, *19*, 97–103.
19. Yuan, Y.F. *Changes in Summer Precipitation and the Impact of Urban Heat Island on Hourly Precipitation Extremes in Beijing Area*; Nanjing University of Information Science and Technology Press: Nanjing, China, 2017.
20. Yang, X.C.; Hou, Y.L.; Chen, B.D. Observed surface warming induced by urbanization in east China. *J. Geophys. Res.* **2011**, *116*, D14113. [CrossRef]
21. Wu, M.J. *Synoptic Situations of Extreme Hourly Precipition over China and Its Observed Link to Urbanization*; University of Chinese Academy of Sciences Press: Beijing, China, 2019.
22. Wu, M.W.; Luo, Y.L.; Chen, F.; Wong, W.K. Observed Link of Extreme Hourly Precipitation Changes to Urbanization over Coastal South China. *J. Appl. Meteorol. Climatol.* **2019**, *58*, 1799–1819. [CrossRef]
23. Wei, F.Y. *Climatological Statistical Diagnosis and Prediction Technology*; Meteorological Press: Beijing, China, 2007; pp. 23–104.
24. Mann, H.B. Nonparametric tests against trend. *Econometrica* **1945**, *13*, 245–259. [CrossRef]
25. He, C.Y.; Shi, P.J.; Li, J.G.; Chen, D.; Pan, Y.Z.; Li, D.; Zhou, L.; Ichinose, T. Restoring urbanization process in China in the 1990s by using non-radian cecalibrated DMSP/OLS nighttime light lmagery and statistical data. *Chin. Sci. Bull.* **2006**, *51*, 856–861. [CrossRef]
26. Shu, S.; Yu, B.L.; Wu, J.P.; Liu, G.X. Methods for deriving urban built-up area using night-lingt data: Assessment and application. *Remote Sens. Tech. Appl.* **2011**, *26*, 169–176. [CrossRef]
27. Zheng, Y.F.; Fu, Y.; Yin, J.F. Impact of urbanization on summer temperature in the middle and lower reaches of Yangtze River in atmospheric circulation background. *J. Trop. Meteor.* **2014**, *30*, 293–301. [CrossRef]
28. Ao, X.; Zhai, Q.F.; Cui, Y. Analysis of urbanization effect on near-surface wind speed change in Liaoning Province, China. *Meteorol. Mon.* **2020**, *46*, 1153–1164.
29. Ao, X.; Cui, Y.; Zhai, Q.F. Effect of urbanization on the temperature trend of Liaoning Province, China. *J. Glaciol. Geocryol.* **2020**, *42*, 1067–1076.
30. Chu, Z.Y.; Ren, G.Y. Change in urban heat island magnitude and its effect on mean air temperature record in Beijing region, China. *Acta. Meteorol. Sin.* **2005**, *63*, 534–540.
31. Xiao, G.L.; Chen, C.L.; Long, X.H.; Chen, G.H.; Cai, K.Z.; Chen, P. Spatiotemporal characteristics of the short-term rainstorm and heavy rainstorm events in Liaoning province. *J. Meteorol. Environ.* **2019**, *35*, 46–52.
32. Yu, R.C.; Zhou, T.J.; Xiong, A.Y.; Zhu, Y.J.; Li, J.M. Diurnal variations of summer precipitation over contiguous China. *Geophys. Res. Lett.* **2007**, *34*, L01704. [CrossRef]
33. Hou, Y.L.; Chen, B.D. Diurnal variations of summer precipitation in Shanghai. *J. Trop. Meteorol.* **2019**, *25*, 414–420. [CrossRef]
34. Liu, W.D.; You, H.L.; Ren, G.Y.; Yang, P.; Zhang, B.Z. AWS precipitation characteristics based on K-means clustering method in Beijing area, China. *Meteorol. Mon.* **2014**, *40*, 844–851. [CrossRef]
35. Lin, X.; Randall, A.D.; Fowler, L.D. Diurnal variability of the hydrologic cycle and radiative fluxes: Comparisons between observations and a GCM. *J. Clim.* **2000**, *13*, 4159–4179. [CrossRef]
36. Nesbitt, S.W.; Zipser, E.J. The diurnal cycle of rainfall and convective intensity according to three years of TRMM measurements. *J. Clim.* **2003**, *15*, 1456–1475. [CrossRef]
37. Li, J.; Yu, R.C.; Wang, J.J. Diurnal variations of summer precipitation in Beijing. *Chin. Sci. Bull.* **2008**, *53*, 1933–1936. [CrossRef]

38. Liu, W.D.; You, H.L.; Ren, G.Y.; Yang, P.; Zhang, B.Z. Subtle precipitation characteristics in Beijing area, China. *Clim. Environ. Res.* **2014**, *19*, 61–68. [CrossRef]
39. Yin, S.Q.; Li, W.J.; Chen, D.L.; Chen, G.H.; Cai, K.Z.; Chen, P. Diurnal variations of summer precipitation in the Beijing area and the possible effect of topography and urbanization. *Adv. Atmos. Sci.* **2019**, *35*, 46–52. [CrossRef]
40. Shepherd, J.M.; Pierce, H.; Negri, A.J. Rainfall modification by major urban areas: Observations from spaceborne rain radar on the TRMM satellite. *J. Appl. Meteorol.* **2002**, *41*, 689–701. [CrossRef]
41. Guo, X.L.; Fu, D.H.; Wang, J. Mesoscale convective precipitation system modified by urbanization in Beijing City. *Atmos. Res.* **2006**, *82*, 112–126. [CrossRef]
42. Wang, J.L.; Zhang, R.H.; Wang, Y.C. Areal differences in diurnal variations in summer precipitation over Beijing metropolitan region. *Theor. Appl. Climatol.* **2012**, *110*, 395–408. [CrossRef]
43. Tan, J.G.; Gu, W. Research Progress on Urban-Induced Rainfall Effect, China. *Adv. Meteorol. Sci. Technol.* **2015**, *5*, 17–22. [CrossRef]
44. Yonaba, R.; Koïta, M.; Mounirou, L.A.; Tazen, F.; Queloz, P.; Biaou, A.C.; Niang, D.; Zouré, C.; Karambiri, H.; Yacouba, H. Spatial and transient modelling of land use/land cover (LULC) dynamics in a Sahelian landscape under semi-arid climate in northern Burkina Faso. *Land Use Policy* **2021**, *103*, 105305. [CrossRef]
45. Zhang, W.F.; Yang, J.C.; Yang, L.; Niyogi, D. Impacts of City Shape on Rainfall in Inland and Coastal Environments. *Earth Future* **2022**, *10*, e2022EF002654. [CrossRef]

land

MDPI

Article

Assessing the Spatial Variability of Daytime/Nighttime Extreme Heat Waves in Beijing under Different Land-Use during 2011–2020

Xiaokang Su [1,2], Fang Wang [2,*], Demin Zhou [1] and Hongwen Zhang [3]

[1] College of Resource Environment and Tourism, Capital Normal University, Beijing 100048, China
[2] Key Laboratory of Land Surface Pattern and Simulation, Institute of Geographic Sciences and Natural Resources Research, Chinese Academy of Sciences, Beijing 100101, China
[3] Beijing Meteorological Disaster Prevention Center, Beijing 100089, China
* Correspondence: wangf@igsnrr.ac.cn; Tel.: +86-010-64889829

Abstract: Urban land-use affects surface air temperature; however, the impact of urban land-use on surface air temperature, particularly the extent to which it affects the duration of extreme heat waves, remains uncertain and the mechanisms of diurnal differences need to be further explored. This paper presents study of daytime/nighttime extreme heat waves duration in Beijing under different land-use changes by adopting an index of cumulative hours of extreme heat waves exceeding the certain thresholds. The urban day/night extreme heat waves cumulative hourly interpolation models were established based on high-resolution urban land-use and socioeconomic data and were assessed to have good performance. The annual average cumulative hours of extreme heat waves increased by 95% (daytime) and 116% (nighttime) in 2016–2020 compared to 2011–2015. The cumulative hours for each land-use type ranked as follows: urban land > cropland > water > grassland > woodland. We found that the cumulative hours of extreme heat waves increased significantly with the proportion of urban land and decreased significantly with the proportion of forested land and water. This research provided important information for alleviating extreme heat waves in cities and for rational land planning.

Keywords: extreme heat waves; land use/cover; interpolation models; daytime; nighttime

1. Introduction

Under the background of global climate change, the frequent occurrence of extremely high temperature weather has a great impact on human health and social economy [1,2]. Studies have shown that land use/cover changes alter the surface roughness, vegetation coverage, and surface reflectivity, which directly affect the latent heat, sensible heat, and water exchange between the surface and the atmosphere, thereby affecting the local regional climate and promoting thermal environment changes [3,4]. Urban land expansion is closely related to urban high temperatures [5–7], and urban land use expansion changes the thermodynamic characteristics of the underlying surfaces of urban areas to a large extent, and also has a certain impact on the global temperature increase [8,9]. Therefore, land-use change can be regarded as the main factor affecting urban high temperatures [10].

Some studies explored the relationship between urban land-use change and urban land surface temperature from different perspectives. Studies were made of the impact of urban expansion on urban land surface high temperatures by analyzing the relationships between various types of building indices or construction land scale and surface temperature [11–13]. Some studies analyzed the relationships between changes of different land-use types and surface temperature [14,15] or usage of remote sensing indices (Normalized Difference Vegetation Index (NDVI), Normalized Difference Water Index (NDWI), etc.) to conduct correlation with surface temperature [16,17]. Some scholars estimated the contribution rates of different land-use types to high temperatures [18–21]. In addition, related research introduced the landscape pattern index to study their impact on urban land surface high temperatures [22,23]. In general, most of the current studies have focused on surface temperature rather than surface air temperature. However, the impact of urban land-use on surface air temperature, especially the extent of the effect on the duration of extreme heat waves, remains uncertain, and the mechanisms of their diurnal differences need to be further explored.

In this study, we explored daytime/nighttime extreme heat waves changes in Beijing under different land-use changes from 2011 to 2020 spanning the 12th Five-Year Plan and 13th Five-Year Plan. Data measured by 225 meteorological stations in the Beijing area is used in this study and are provided by the Beijing Meteorological Disaster Prevention Center. The 90% quantile was taken as the high temperature threshold, and the cumulative hours of extreme heat waves exceeding the thresholds (daytime: 33.1 °C; nighttime: 27.9 °C) were adopted to characterize the intensity and duration of extreme heat waves. We established the urban day/night extreme heat waves interpolation models based on high-resolution urban land use and socioeconomic data during 2011–2020. The relationship between different land-use changes and accumulated hours of day/night extreme heat waves was explored.

2. Materials and Methods

2.1. Study Area

Beijing is located within the range of 115.7° E–117.4° E, 39.4° N–41.6° N, with an altitude range of 6–2300 m. It has a jurisdiction with over 16 districts and 331 townships, covering an area of 16,410.54 km^2 (Figure 1a). At the end of 2020, Beijing's permanent population was 21.893 million of which the urban population was 19.166 million, accounting for 87.5%. The gross domestic product (GDP) of Beijing is 3.6 trillion yuan, and the city is worthy of being called an international metropolis and representative of China's rapid urbanization (Figure 1b). The region has a typical temperate sub-humid continental monsoon climate with high temperatures and a rainy summer. Through calculation, we found that, from 2011 to 2020, the average temperature in summer (June–August) in Beijing increased at a rate of 0.64 °C/10a, and the number of hot days also increased, making Beijing one of the hottest cities in China.

2.2. Data Collection and Preprocessing

2.2.1. Meteorological Data

According to the official data published by the China Meteorological Administration, the average daily maximum temperature in Beijing was significantly higher in June, July, and August than in other months. Therefore, in this study, summer was chosen as the time period when extreme heat waves occur frequently in Beijing. Hourly temperature data from Beijing meteorological stations (MSs) from June to August during 2011–2020, provided by the National Meteorological Information Center, were used in this study. We selected a total of 225 MSs (20 national weather stations-NWSs, 205 automatic weather stations-AWSs) that were established before June 2011 and still in use until September 2020 (Figure 1a). Most NWSs are responsible for regional or national weather information exchange and are the main body of the national weather and climate website. The observation data obtained by AWS are mainly used for weather services in their own provinces (districts and cities) and localities and also complement the observation data on the national weather and climate website. All observation data were processed by the National Meteorological Information Center for strict quality control. The outliers in the original data were eliminated, and the missing data are replaced by observations at adjacent times or average values, reducing the error caused by instrument failure or measurement error, and the data are highly accurate.

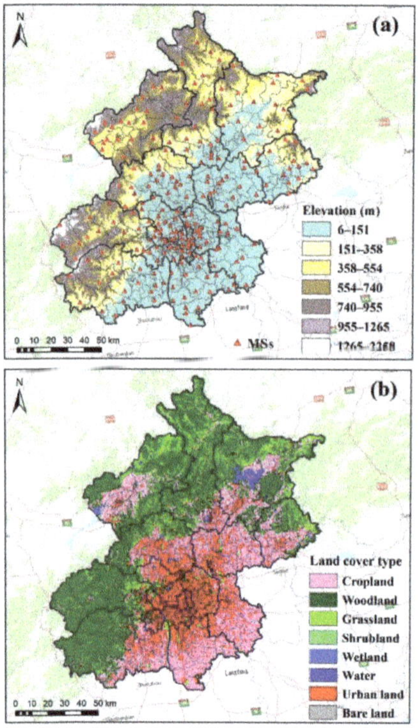

Figure 1. (**a**) Elevations and locations of meteorological stations (MSs). (**b**) Land cover types of cropland, woodland, grassland, shrubland, wetland, water, urban land, and bare land in 2020 (The black border represents the administrative border of the district; the gray border represents the administrative border of the township. Data on administrative boundaries were obtained from the National Geomatics Center of China).

2.2.2. Land Cover

Land cover data were from the GlobeLand30 V2010 and GlobeLand30 V2020 database (downloaded at http://www.globallandcover.com (accessed on 8 July 2022)), with a spatial resolution of 30 m, include 10 types, namely: cropland, woodland, grassland, shrubland, wetland, water, tundra, urban land, bare land, and glacier. The data accuracy of GlobeLand30 V2010 was evaluated by Tongji University. Over 150,000 test samples were laid down, and the overall accuracy of GlobeLand30 V2010 data was 83.50% with a Kappa coefficient of 0.78. The data accuracy of GlobeLand30 V2020 was evaluated by the Aerospace Information Research Institute, Chinese Academy of Sciences. With over 230,000 samples laid down, the overall accuracy of GlobeLand30 V2020 data was 85.72% with a Kappa coefficient of 0.82. The Kappa coefficient is used for consistency testing and can also be used to measure classification accuracy. In general, a higher Kappa coefficient indicates a higher classification accuracy.

2.2.3. Population Density

The spatial distribution kilometer grid data of China's population density in 2015 and 2020 were used, which were from the Institute of Geographic Sciences and Natural Resources Research (IGSNRR, https://www.resdc.cn/ (accessed on 8 July 2022)), with a resolution of 1 km. The population density unit is person/km^2. The cumulative population of each district in Beijing was calculated by the ArcGIS statistical tool and verified by the resident population of each district in Statistical Yearbook data. The two showed a strong correlation, with R^2 values of 0.99 (2015) and 0.99 (2020).

2.2.4. GDP

Spatial distribution kilometer grid data of China's GDP in 2020 from the IGSNRR were used, with a resolution of 1 km. Each grid cell represents the total GDP of the area in units of 10,000 yuan/km^2. The cumulative GDP of each district in Beijing was calculated using the ArcGIS zoning statistical tool and verified using the total GDP of each district in the Beijing Regional Statistical Yearbook 2021 (http://nj.tjj.beijing.gov.cn/nj/qxnj/2021/zk/indexch.htm (accessed on 8 July 2022)). The two showed a strong correlation with R^2 values of 0.96 (2015) and 0.96 (2020).

2.2.5. DEM

Digital Elevation Model (DEM) data were derived from the Shuttle Radar Topography Mission (SRTM) data of the US Space Shuttle Endeavor. In this study, the latest SRTM V4.1 data were resampled to generate a new dataset with a resolution of 250 m. Data were projected using a WGS84 ellipsoid.

2.3. *Quantification of Extreme Heat Waves and Spatial Mapping*

In this study, we used the hourly air temperature data of 20 national weather stations and 205 automatic weather stations for the past 10 years (2011–2020). This is the expanded dataset with the largest number of stations and the most abundant data. Based on hourly temperature data, the daytime (6:00–19:00) and nighttime (19:00–6:00) data in the past 10 years were sorted from small to large, and the 90% quantile was taken as the high temperature threshold. The cumulative hours of extreme heat waves at each station exceeding 33.1 °C during the daytime, during the summer (June to August), from 2011 to 2015 (12th Five-Year Plan) or from 2016 to 2020 (13th Five-Year Plan), were calculated, referred to as EHW_{125d} and EHW_{135d}. The cumulative number of hours over 27.9 °C at night at each station was calculated, referred to as EHW_{125n} and EHW_{135n}. Land cover (cropland, woodland, grassland, shrubland, wetland, water, urban land, and bare land), population density, DEM, and GDP, as the predictors, which affected the spatial distribution of high temperature, were selected [24,25]. Then, the extreme heat waves interpolation models were established based on urban surface land features as follows:

$$EHW_n = f(DEM, land_n, pop, GDP) \quad (1)$$

where DEM is the elevation of the weather station, $Land_n$ represents the area of each land use type within the optimal buffer width of the land use type, pop is the population density of the site's location, GDP represents the gross domestic product, EHW_n represents EHW_{125d}, EHW_{125n}, EHW_{135d}, or EHW_{135n}.

To obtain the spatial scale with the strongest correlations between various factors and EHW_n, 225 MSs were set with 100 buffer widths ranging from 1 to 100 km, and the land cover grid data were cropped with the vector files of the buffer. Using the spatial analysis function of ArcGIS, areas of 8 land cover types in 100 buffer zones of approximately 225 MSs were obtained (8 × 100 = 800 variables). DEM, GDP and population density data were point data (3 variables). In summary, 803 predictor variables were prepared.

The Pearson correlation coefficient is used to calculate the type of linear relationship between two variables (positive, negative, none) and the strength of this relationship (weak, moderate, strong). We used Pearson correlation coefficient to analyze the correlation between each variable and EHWn. Since the correlation coefficients between shrubland, wetland and water areas within the 1–100 km buffer zone and EHW_n were not high, these variables were removed. Then, the buffer width with the highest correlation coefficient with EHW_n was determined among the 100 buffer widths for the 4 land types of cropland, woodland, grassland and urban land. Stepwise linear regression was performed with EHW_n as dependent variables, and cropland, woodland, grassland, urban land, population density,

DEM, and GDP as independent variables. Then, the multiple linear regression equations are obtained, which were the extreme heat waves interpolation models of EHW_{125d}, EHW_{125n}, EHW_{135d}, and EHW_{135n} (see Supplementary Materials).

Based on the above multiple regression models, we interpolated the EHW_n data of the kilometer grid cells using daytime and nighttime data from 225 stations, combined with resolution land use data of 30 m. First, 1 km × 1 km regular grid cells and grid points were generated for the study area; we calculated the value of the independent variable and the value of the result variable for each grid point and then we assigned the values of grid points to the grid to obtain the spatial distribution maps of daily/night extreme heat waves during the 12th and 13th Five-Year plans with a resolution of 1 km.

3. Result Analysis

3.1. Performance Evaluation of the Interpolation Models for Cumulative Hours of Urban Extreme Heat Waves

We used EHW_{125d}, EHW_{125n}, EHW_{135d}, and EHW_{135n} from 225 MSs to correlate with the area of each land-use type within the 1–100 km buffer zone (Figure 2). With the change of buffer width, the cropland area in the buffer zone was positively correlated with EHW_{125d}, EHW_{125n}, EHW_{135d}, and EHW_{135n}. When the buffer width was 0–50 km, the absolute value of the correlation coefficient was less than 0.4, which was a weak correlation. When the buffer width was 50–100 km, the absolute value of the correlation coefficient was greater than 0.4 and less than 0.6, which was a moderate correlation. The area of woodland in the buffer zone was negatively correlated with EHW_{125d}, EHW_{125n}, EHW_{135d}, and EHW_{135n}. When the buffer width was 0–60 km, the absolute value of the correlation coefficient was greater than 0.6 and less than 0.8, which was a strong correlation. When the buffer width was 60–100 km, the absolute value of the correlation coefficient is greater than 0.4 and less than 0.6, which was a moderate correlation. The area of grassland in the buffer zone was negatively correlated with EHW_{125d}, EHW_{125n}, EHW_{135d}, and EHW_{135n}. When the buffer width was 0–40 km, the absolute value of the correlation coefficient was greater than 0.4 and less than 0.6, which was a moderate correlation. When the buffer width was 40–100 km, the absolute value of the correlation coefficient was greater than 0.6 and less than 0.8, which was a strong correlation. The area of urban land in the buffer zone was positively correlated with EHW_{125d}, EHW_{125n}, EHW_{135d}, and EHW_{135n}, and the absolute value of the correlation coefficient was greater than 0.6, which was a strong correlation. The area of shrubs, wetlands, and water in the buffer zone and EHW_{125d}, EHW_{125n}, EHW_{135d}, and EHW_{135n} were sometimes positively correlated and sometimes negatively correlated, which may be related to the area and distribution pattern of shrubs, wetlands, and waters in Beijing.

By stepwise multiple regression, we obtained the multiple linear regression equations for EHW_{125d}, EHW_{125n}, EHW_{135d}, and EHW_{135n} (see Supplementary Materials), which showed that urban land and DEM were predictors that enter into each resulting equation. This indicated that urban land and DEM were the variables most associated with EHW_{125d}, EHW_{125n}, EHW_{135d}, and EHW_{135n} compared to other variables. The significance values of the four multiple linear regression equations by F-test ANOVA are all less than 0.01, indicating that the equations are highly significant. R^2 are 0.72 (EHW_{125d}), 0.77 (EHW_{125n}), 0.73 (EHW_{135d}), and 0.74 (EHW_{135n}), respectively, indicating that the fitting degree of the model is good. The fitting quality for nighttime (EHW_{125n} and EHW_{135n}) is better than that for daytime (EHW_{125d} and EHW_{135d}).

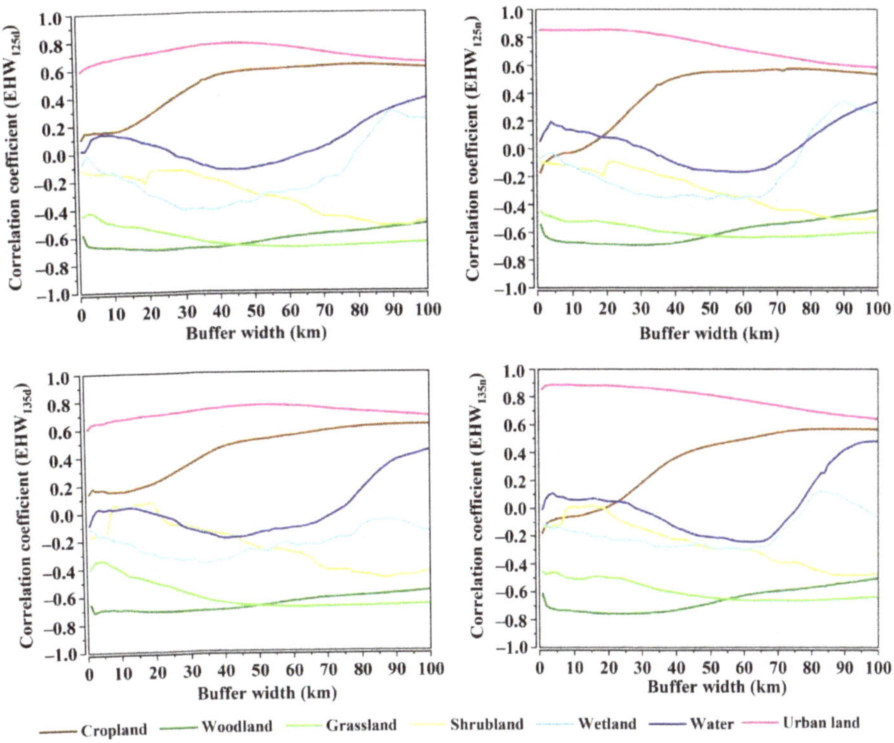

Figure 2. Correlation coefficients between the area of each land-use type in the 1–100 km buffer zones of the 225 MSs with EHW_{125d}, EHW_{125n}, EHW_{135d}, and EHW_{135n}. (EHW_{125d} represents the cumulative hours of daytime extreme heat waves during the 12th Five-Year Plan period, EHW_{125n} represents the cumulative hours of nighttime extreme heat waves during the 12th Five-Year Plan period, EHW_{135d} represents the cumulative hours of daytime extreme heat waves during the 13th Five-Year Plan period, and EHW_{135n} represents the cumulative hours of nighttime extreme heat waves during the 13th Five-Year Plan period).

3.2. *Spatial Pattern Analysis of Cumulative Hours of Extreme Heat Waves during the Day/Night*

The spatial distributions of EHW_{125d}, EHW_{125n}, EHW_{135d}, and EHW_{135n} all show the pattern "high inside and low outside; high in the south and low in the north" (Figure 3), with a circle-layer structure that diverges from the central city to the suburbs. We resampled the raster map with a resolution of 1 km to the township scale and calculated the cumulative hours of extreme heat waves for each township. Then, divided the accumulated hours of extreme heat waves into 10 levels (Figure 3), where levels 1, 2, and 3 are low levels; levels 4, 5, and 6 are medium levels; levels 7, 8, and 9 are high levels; and level 10 is very high. Counted the number of townships that belong to the cumulative hours of extreme heat waves at different levels (Table 1). In 2011–2015, during the day, 68 townships were at low levels, 236 towns were at medium levels, and 27 towns were at high levels. At night, 94 townships were at low levels, 107 townships were at medium levels, and 130 townships were at high levels. In 2016–2020, during the day, 33 townships were at low levels, 35 towns were at medium levels, 102 towns were at high levels, and 161 towns were at the very high level. At night, 33 townships were at low levels, 38 townships were at medium levels, 129 townships were at high levels, and 131 townships were at the very high level. Compared with 2011–2015, from 2016 to 2020, there were 95 more high-level townships and 161 more extremely high-level townships in daytime, one less high-level townships and 131 more extremely high-level townships at night. This is because temperatures generally increased between 2011 and 2020, and daytime temperatures increased even more. This is due to the general increase in temperature between 2011 and 2020, so the number of townships that are at extremely high-level during the daytime and nighttime has increased. In addition, the temperature rises more during the daytime compared to the nighttime, so the number of townships at extremely high-level during the daytime increases more.

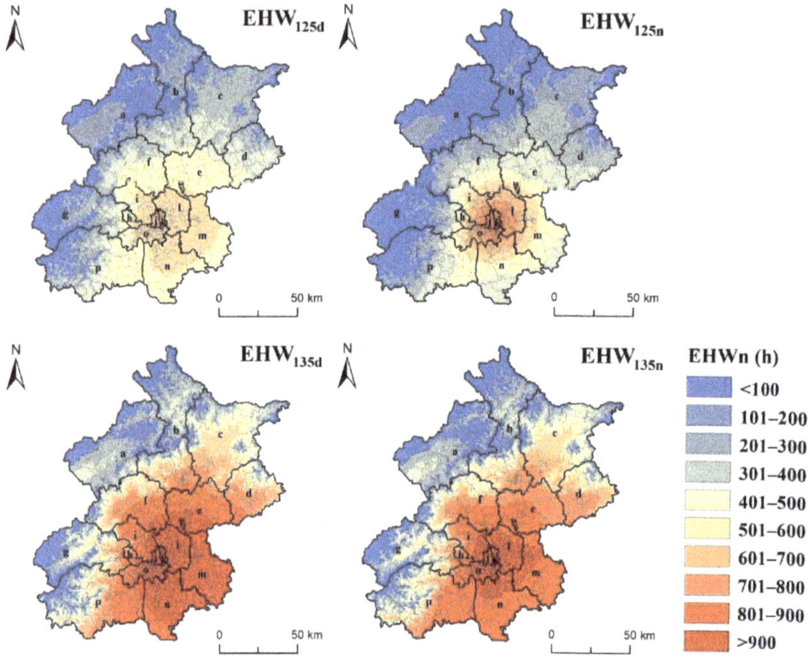

Figure 3. Spatial distribution maps of EHW_{125d}, EHW_{125n}, EHW_{135d}, and EHW_{135n} in Beijing with a resolution of 1 km. (EHWn represents EHW_{125d}, EHW_{125n}, EHW_{135d}, or EHW_{135n}).

Table 1. The number of townships that belong to the cumulative hours of extreme heat waves at different levels.

	EHW_{125d}	EHW_{125n}	EHW_{135d}	EHW_{135n}
Level 1, 2, 3	68	94	33	33
Level 4, 5, 6	236	107	35	38
Level 7, 8, 9	27	130	102	129
Level 10	-	-	161	131

3.3. Comparison of Accumulated Hours of Day/Night Extreme Heat Waves for Different Land-Use Types

The average values of EHW_{125d}, EHW_{125n}, EHW_{135d}, and EHW_{135n} on the surface of each land-use type in the whole Beijing, Beijing's urban areas, and Beijing's suburbs, were obtained through the spatial overlay of the land-use type vector map with EHW_{125d}, EHW_{125n}, EHW_{135d}, and EHW_{135n}. It can be seen in Figure 4 that the order of the average of cumulative hours of extreme heat waves on the surface of each land-use type in Beijing during the 12th and 13th Five-Year Plans was: urban land > cropland > water > grassland > woodland. During the 12th and 13th Five-Year Plan period, the order of the average of cumulative hours of extreme heat waves on the surface of each land-use type in Beijing's suburbs was as follows: urban land > cropland > water area > grassland > woodland. During the 12th Five-Year

Plan period, the order of the average of accumulated hours of extreme heat waves on the surface of each land-use type in the urban area of Beijing during the day was as follows: urban land > water > woodland > cropland > grassland. Additionally, the order of the average of accumulated hours of extreme heat waves on the surface of various land-use types in urban areas of Beijing at night was as follows: urban land > water > cropland > woodland > grassland. During the 13th Five-Year Plan period, the order of the average of accumulated hours of extreme heat waves on the surface of each land-use type in the urban area of Beijing during the day was as follows: urban land > water > cropland > grassland > woodland; the order of the average of cumulative hours of extreme heat waves corresponding to the surface of each land-use type in Beijing urban area at night was: urban land > water > cropland > woodland > grassland.

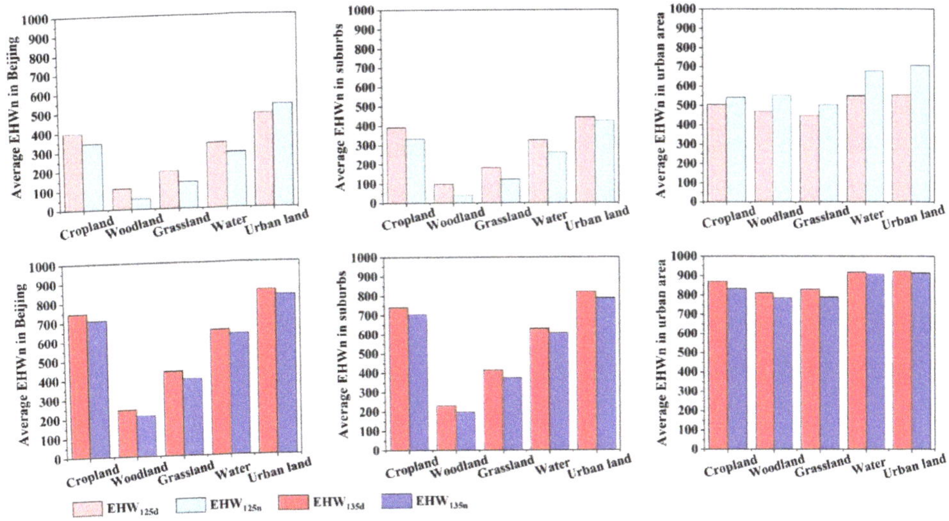

Figure 4. The average values of EHW_{125d}, EHW_{125n}, EHW_{135d}, and EHW_{135n} for each land-use type in the whole Beijing, Beijing's urban areas, and Beijing's suburbs.

3.4. *Influence of Different Spatial Combination Patterns of Land-Use Type on the Cumulative Hours of Day/Night Extreme Heat Waves*

In order to further explore the impact of the spatial pattern of land-use types on the cumulative hours of day/night extreme heat waves, the land use pattern was represented by the proportions of the area occupied by the land-use types within a 1 km grid, and we analyzed their relationships with EHW_{125d}, EHW_{125n}, EHW_{135d}, and EHW_{135n}. First, the area proportion of each land-use type in each grid and the values of EHW_{125d}, EHW_{125n}, EHW_{135d}, and EHW_{135n} in each grid were calculated. Then, the mean values of EHW_{125d}, EHW_{125n}, EHW_{135d}, and EHW_{135n} of each land-use type at various proportions in the grid were counted at intervals of 20% for comparative analysis. From the change curves of the average values of EHW_{125d}, EHW_{125n}, EHW_{135d}, and EHW_{135n} under different proportions of each land-use type (Figure 5), it can be seen that during the 12th and 13th Five-Year Plans, with the increase in the proportion of cropland in the grid, the average values of EHW_{125d}, EHW_{125n}, EHW_{135d}, and EHW_{135n} of the grid also increased. Additionally, the increase rate during the day was greater than that at night, and the increase rate during the 13th Five-Year Plan period was greater than that during the 12th Five-Year Plan period. During the 12th and 13th Five-Year Plan period, with the increase in the proportion of woodland in the grid, the average values of EHW_{125d}, EHW_{125n}, EHW_{135d}, and EHW_{135n} of the grid decreased, and the rate of decrease at night was greater than that during the day. The rate of decrease in the five-year period was greater than that in the 12th five-year period. During the 12th Five-Year Plan and the 13th Five-Year Plan (when the area proportion of water in the grid was greater than 40%), as the proportion of water in the grid increased, the average values of EHW_{125d}, EHW_{125n}, EHW_{135d}, and EHW_{135n} of the grid increased. During the 12th Five-Year Plan period, the rate of decrease at night was greater than that during the day, and during the 13th Five-Year Plan period, the rate of decrease during the day was greater than that at night. During the 12th and 13th Five-Year Plans, as the proportion of urban land in the grid increased, the average EHW_{125d}, EHW_{125n}, EHW_{135d}, and EHW_{135n} of the grid also increased. Additionally, the increase rate at night was greater than that during the day, and the increase rate during the 13th Five-Year Plan period was greater than that during the 12th Five-Year Plan period.

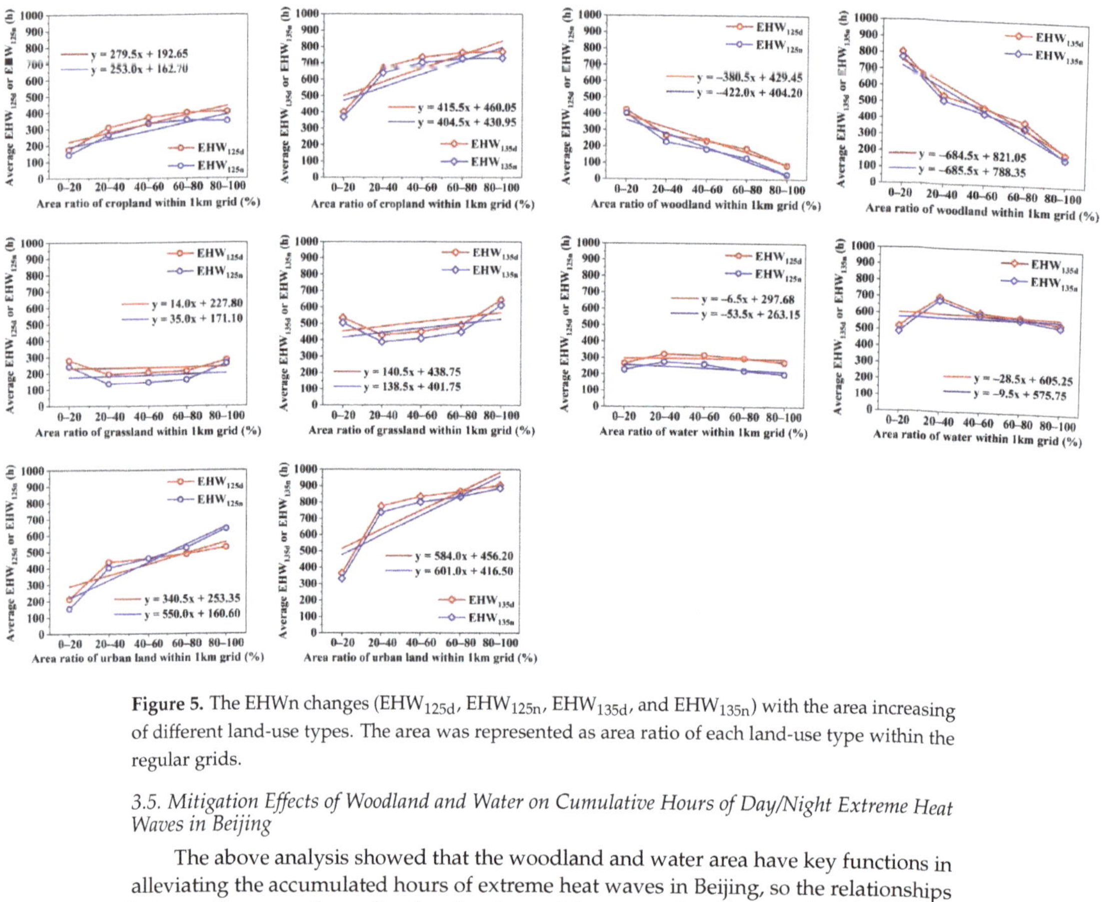

Figure 5. The EHWn changes (EHW_{125d}, EHW_{125n}, EHW_{135d}, and EHW_{135n}) with the area increasing of different land-use types. The area was represented as area ratio of each land-use type within the regular grids.

3.5. Mitigation Effects of Woodland and Water on Cumulative Hours of Day/Night Extreme Heat Waves in Beijing

The above analysis showed that the woodland and water area have key functions in alleviating the accumulated hours of extreme heat waves in Beijing, so the relationships between the areas of woodland and water and the accumulated hours of urban extreme heat waves were further analyzed. The grid-by-grid statistical analysis results of EHW_{125d}, EHW_{125n}, EHW_{135d}, and EHW_{135n} (Table 2) show that the grids without woodland and water had the highest average values of EHWn, which were 444 (EHW_{125d}), 418 (EHW_{125n}), 849 (EHW_{135d}), and 812 (EHW_{135n}), respectively. The second highest average values of EHWn were for grids that were entirely water areas: 277 (EHW_{125d}), 202 (EHW_{125n}), 548 (EHW_{135d}), and 533 (EHW_{135n}). The average values of EHWn of the grids that were completely woodland were the lowest, which were 33 (EHW_{125d}), 0 (EHW_{125n}), 97 (EHW_{135d}), and 68 (EHW_{135n}). The average values of EHWn of the grids with woodland proportions greater than 50% were 114 (EHW_{125d}), 57 (EHW_{125n}), 251 (EHW_{135d}), and 218 (EHW_{135n}), and the average values of EHWn of grids with water proportions greater than 50% were 291 (EHW_{125d}), 226 (EHW_{125n}), 565 (EHW_{135d}), and 549 (EHW_{135n}). It can be found that, regardless of the spatial distribution of woodland and water, the average values of EHW_{125n} are smaller than that of EHW_{125d}, and the average values of EHW_{135n} is smaller than that of EHW_{135d}. It can be seen that the woodland and water areas have obvious effects on alleviating the accumulated hours of extreme heat waves, and the mitigation effect at night is stronger than that in the daytime, and the mitigation effect of woodland is stronger than that of water area.

Table 2. Average values of EHW_{125d}, EHW_{125n}, EHW_{135d}, and EHW_{135n} under different proportions of woodland and water area.

	EHW_{125d} (h)	EHW_{125n} (h)	EHW_{135d} (h)	EHW_{135n} (h)
Woodland (0%) & Water (0%)	444	418	849	812
Woodland (100%)	33	0	97	68
Water (100%)	277	202	548	533
Woodland (>50%)	114	57	251	218
Water (>50%)	291	226	565	549

4. Discussion

This study analyzed daytime and nighttime cumulative hours of extreme heat waves under different land-use types in the representative metropolis of Beijing, and gave the high-resolution spatial mapping of day/night extreme heat waves cumulative hours. This study further explored influence of different land-use patterns on accumulated hours of day/night extreme heat waves. The results provided important reference for alleviating extreme heat waves in cities and for rational land planning. When cities expand rapidly, in addition to considering the expansion of building sites (due to their potential for extreme heat waves), other land use types (such as woodlands, grasslands, and water) should be considered to mitigate extreme heat waves.

Based on land use data with a resolution of 30 m and socioeconomic data with a resolution of 1km, this study established the urban day/night extreme heat waves interpolation model. Both the significance test and R^2 analysis showed that the model has good performance. The interpolation quality for nighttime was better than that for daytime. It can be a reference method for interpolating temperature data. In the final model, only the two to three most influential and contributing predictors were selected from a broad set of candidate predictors, mainly including DEM, population density, area of grassland and area of urban land within a specific buffer width. This suggests that DEM, population density, area of grassland, and urban land have the greatest correlation on urban extreme heat waves. It is worth noting that the land use patterns in different regions are different, so the obtained multiple regression interpolation models will also be different.

The spatial distribution of the cumulative hours of day/night extreme heat waves in Beijing is "high inside and low outside; high in the south and low in the north". In other words, the cumulative hours of extreme heat waves in the eight districts of Dongcheng, Xicheng, Chaoyang, Haidian, Fengtai, Shijingshan, Daxing, and Tongzhou were relatively high, while the cumulative hours of extreme heat waves in the remaining eight districts were relatively low. This is related to factors, such as land use, population density, DEM, etc. The area ratio of urban land in the eight districts of Dongcheng, Xicheng, Chaoyang, Haidian, Fengtai, Shijingshan, Daxing, and Tongzhou is 54.22%, the area ratio of woodland is 5.15%, the average population density is 4301 persons/km^2, and the average altitude is 44.83m. In the other eight districts, the proportion of woodland is 55.20%, the proportion of urban land is 12.84%, the average population density is 518 persons/km^2, and the average altitude is 450.43 m. These differences are important reasons for this spatial distribution pattern.

We simulated the spatial distribution pattern of accumulated hours of day/night extreme heat waves in Beijing in 2011–2015 and 2016–2020. In 2016–2020, the average annual cumulative hours of extreme heat waves were 95.06% (daytime) and 115.70%

(nighttime) higher than those in 2011–2015. This is related to natural climate change and human activities [26–28]. The WMO survey results show that 2011–2020 was the warmest decade on record. The warmest six years were 2015 and subsequent years, with 2016, 2019, and 2020 making up the top three. This is an important reason why the accumulated hours of extreme heat waves in 2015–2020 were much higher. Compared with 2011–2015, the average annual resident population in 2016–2020 increased by 3.54%, and the GDP increased by 64.41%. Compared with 2010, the urban land area by 2020 had increased by 55.43%. Human activities, such as population movement, urban land expansion, and fossil fuel burning, are also another important reason why the cumulative hours of extreme heat waves in 2016–2020 were much more than those in 2011–2015.

Our research shows that the order of land-use types based on the cumulative hours of extreme heat waves from different land types in Beijing were as follows: urban land > cropland > water > grassland > woodland. Urban land had the most cumulative hours of extreme heatwaves during the day and night, the woodland had the fewest cumulative hours of extreme heat waves during the day and night. Cropland and urban land can increase the cumulative hours of extreme heat waves, whereas woodland and water areas can reduce the cumulative hours of extreme heat waves. The effect of grassland on the cumulative hours of extreme heat waves was not significant. The role of cropland in enhancing the cumulative hours of extreme heat waves is greater during the day than at night; the role of urban land in enhancing the cumulative hours of extreme heat waves is greater at night than in the daytime, and the role of woodland in alleviating the cumulative hours of extreme heat waves is greater at night than in the daytime. This day–night difference is related to the complex surrounding environment and the differences in day–night specific heat capacities of the various land-use types [29–31]. Water has the highest specific heat capacity, followed by woodlands and grasslands with higher water content (woodlands have higher water content than grasslands) and least impermeable surfaces. Urbanization leads to reduced evaporation and wind. In addition, the lower albedo of impervious surface compared to vegetation increases daytime heat storage and enhances nighttime long-wave heat release. It should be noted that our research shows that grassland has the effect of enhancing the cumulative hours of extreme heat waves. As can be seen in the Figure 5, this result was mainly caused by the high cumulative hours of extreme heat waves in grids with grassland accounting for more than 80% of the area. Both 2010 and 2020, among the 15,818 grids in Beijing, there were only approximately 80 grids that had more than 80% grassland. Nearly 2/3 of the 80 grids were close to cultivated land or urban land. Environmental impacts lead to higher cumulative hours of extreme heat waves. It can be seen in the Figure 5 that the cumulative hours of extreme heat waves for grids with water areas less than 20% were significantly lower than those grids with larger water areas. This is because the water area of Beijing is too small, only 0.92% (2010) or 1.45% (2020). As a result, in 2010 and 2020, the proportions of grids with water area accounting for 0–20% were extremely high, 98.8% and 98.1%, respectively. The grids with water areas of 0–20% are numerous and continuously distributed, making them less susceptible to the surrounding environment. On the contrary, grids with water areas greater than 40% are small in number and scattered, and are easily affected by the surrounding environment, which makes the cumulative hours of extreme heat waves relatively high. However, when the proportion of the water area is greater than 40%, the cumulative hours of extreme heat waves is decreased significantly as the proportion of the water area continues to increase, which reflects the role of water area in relieving the cumulative hours of extreme heat waves. Different from water, the area of woodland in Beijing accounted for 44.74% (2010) and 45.08% (2020), and the distribution was concentrated. This difference between woodland and water in area and spatial pattern may lead to Beijing's woodland having a stronger mitigation effect on the cumulative hours of extreme heat waves relative to water.

5. Conclusions

The main conclusions of our study are summarized as follows:

1. The urban day/night extreme heat waves cumulative hourly interpolation models were established; the correlations were highly significant at $p < 0.01$, both the significance test and R^2 analysis showed that the models have good performance, and the accuracy of interpolation was high;
2. The spatial distribution of the cumulative hours of day/night extreme heat waves in Beijing is "high inside and low outside; high in the south and low in the north"; a circle structure radiates from the central city to the suburbs. During 2016–2020, the annual cumulative hours of extreme heat waves in the daytime and in the nighttime were 95.06% and 115.70% higher than those of 2011–2015;
3. The order of land-use types for cumulative hours of land-surface extreme heat waves in Beijing was as follows: urban land > cropland > water > grassland > woodland. In most cases, the cumulative hours of extreme heatwaves during the day were greater than those at night for cropland, woodland, grassland, water, and urban land;
4. The increase associated with urban land in the cumulative hours of extreme heat waves at night is greater than that in the daytime. The mitigation effects of woodland and water on the cumulative hours of extreme heat waves are stronger at night than in the daytime, and the mitigation effect of woodland is stronger than that of water.

Supplementary Materials: The following supporting information can be downloaded at: https://www.mdpi.com/article/10.3390/land11101786/s1.

Author Contributions: Conceptualization, F.W.; data curation, X.S. and H.Z.; formal analysis, X.S.; funding acquisition, F.W.; methodology, X.S. and F.W.; software, X.S.; supervision, F.W. and D.Z.; writing—original draft, X.S.; writing—review and editing, F.W. All authors have read and agreed to the published version of the manuscript.

Funding: This work is jointly supported by National Key Research and Development Program of China (2019YFC0507805) and the Strategic Leading Science and Technology Program of the Chinese Academy of Sciences (XDA20020202).

Institutional Review Board Statement: Not applicable.

Informed Consent Statement: Not applicable.

Data Availability Statement: The processed data, which were used to generate the figures and tables, are available upon request to the corresponding author.

Acknowledgments: We express our sincere appreciation to the anonymous reviewer for constructive comments.

Conflicts of Interest: The authors declare no conflict of interest. The funders had no role in the design of the study; in the collection, analyses, or interpretation of data; in the writing of the manuscript, or in the decision to publish the results.

References

1. Luber, G.; McGeehin, M. Climate change and extreme heat events. *Am. J. Prev. Med.* **2008**, *35*, 429–435. [CrossRef] [PubMed]
2. Patz, J.A.; Campbell-Lendrum, D.; Holloway, T.; Foley, J.A. Impact of regional climate change on human health. *Nature* **2005**, *438*, 310–317. [CrossRef] [PubMed]
3. Bonan, G.B. Effects of Land Use on the Climate of the United States. *Clim. Chang.* **1997**, *37*, 449–486. [CrossRef]
4. Findell, K.L.; Shevliakova, E.; Milly, P.C.D.; Stouffer, R.J. Modeled impact of anthropogenic land cover change on climate. *J. Clim.* **2007**, *20*, 3621–3634. [CrossRef]
5. Ge, R.F.; Wang, J.L.; Zhang, L.X. Impacts of urbanization on the urban thermal environment in Beijing. *Acta Ecol. Sin.* **2016**, *36*, 6040–6049. (In Chinese)
6. Hu, A.G. Urbanization as an engine of China's economic growth. *Popul. Sci. China* **2003**, *2003*, 5–12. (In Chinese)
7. Feng, Y.Y.; Hu, T.G.; Zhang, L.X. Impacts of structure characterisitics on the thermal environment effect of city parks. *Acta Ecol. Sin.* **2014**, *34*, 3179–3187. (In Chinese)

8. Li, X.B. International research on environmental consequence of land use/cover change. *Adv. Earth Sci.* **1999**, *14*, 395–400. (In Chinese)
9. Stathopoulou, M.; Cartalis, C. Daytime urban heat islands from Landsat ETM+and Corine land cover data: An application to major cities in Greece. *Sol. Energy* **2007**, *81*, 358–368. [CrossRef]
10. Yue, W.Z.; Xu, L.H. Thermal environment effect of urban land use type and pattern—A case study of central area of Shanghai city. *Sci. Geogr. Sin.* **2007**, *27*, 243–248. (In Chinese)
11. Polydoros, A.; Cartalis, C. Assessing the impact of urban expansion to the state of thermal environment of peri-urban areas using indices. *Urban Clim.* **2015**, *14*, 166–175. [CrossRef]
12. Sharma, R.; Chakraborty, A.; Joshi, P.K. Geospatial quantification and analysis of environmental changes in urbanizing city of Kolkata (India). *Environ. Monit. Assess.* **2015**, *187*, 4206. [CrossRef] [PubMed]
13. Rasul, A.; Balzter, H.; Smith, C.L. Applying a normalized ratio scale technique to assess influences of urban expansion on land surface temperature of the semi-arid city of Erbil. *Int. J. Remote Sens.* **2017**, *38*, 3960–3980. [CrossRef]
14. Zhang, X.L.; Zhang, S.W.; Li, Y. Correlation between urban thermal environment effect and land use pattern in Changchun. *Resour. Sci.* **2008**, *30*, 1564–1570.
15. Guo, A.; Yang, J.; Sun, W.; Xiao, X.; Cecilia, J.X.; Jin, C.; Li, X. Impact of urban morphology and landscape characteristics on spatiotemporal heterogeneity of land surface temperature. *Sustain. Cities Soc.* **2020**, *63*, 102443. [CrossRef]
16. Guo, G.; Wu, Z.; Xiao, R. Impacts of urban biophysical composition on land surface temperature in urban heat island clusters. *Landsc. Urban Plan.* **2015**, *135*, 1–10. [CrossRef]
17. Estoque, R.C.; Murayama, Y.; Myint, S.W. Effects of landscape composition and pattern on land surface temperature: An urban heat island study in the megacities of Southeast Asia. *Sci. Total Environ.* **2017**, *577*, 349–359. [CrossRef]
18. Zhang, L.J.; Chen, H.; Zhang, J.F. Study on the relationship between urban expansion and land surface thermal environment change in Harbin City. *Areal Res. Dev.* **2010**, *29*, 49–52. (In Chinese)
19. Sun, J.; Zhang, H.; Whang, Q. Correlations between land cover type and surface temperature of the Taihu Lake basin based on HJ-1 satellite data. *Res. Environ. Sci.* **2011**, *24*, 1291–1296. (In Chinese)
20. Zhao, Z.Q.; Li, L.G.; Wang, H.B. Study on the relationships between land use and land surface temperature in Shenyang urban districts. *J. Meteorol. Environ.* **2016**, *32*, 102–108. (In Chinese)
21. Yaung, K.L.; Chidthaisong, A.; Limsakul, A.; Varnakovida, P.; Nguyen, C.T. Land Use Land Cover Changes and Their Effects on Surface Air Temperature in Myanmar and Thailand. *Sustainability* **2021**, *13*, 10942. [CrossRef]
22. Connors, J.P.; Galletti, C.S.; Chow, W.T.L. Landscape configuration and urban heat island effects: Assessing the relationship between landscape characteristics and land surface temperature in Phoenix, Arizona. *Landsc. Ecol.* **2013**, *28*, 271–283. [CrossRef]
23. Asgarian, A.; Amiri, B.J.; Sakieh, Y. Assessing the effect of green cover spatial patterns on urban land surface temperature using landscape metrics approach. *Urban Ecosyst.* **2015**, *18*, 209–222. [CrossRef]
24. Yuan, S.; Ren, C.; Meng, C.; Lau, K.K.; Lee, T.C.; Wong, W.K. Assessing spatial variability of extreme hot weather conditions in Hong Kong: A land use regression approach. *Environ. Res.* **2019**, *171*, 403–415.
25. Hua, J.Y.; Zhang, X.Y.; Ren, C.; Yuan, S.; Lee, T.-C. Spatiotemporal assessment of extreme heat risk for high-density cities: A case study of Hong Kong from 2006 to 2016. *Sustain. Cities Soc.* **2021**, *64*, 102507. [CrossRef]
26. Yao, Y.; Chen, X.; Qian, J. The research progress of urban surface thermal environment. *Acta Ecol. Sin.* **2018**, *38*, 1134–1147. (In Chinese)
27. Oke, T.R.; Mills, G.; Christen, A. *Urban Climates*; Miao, S.; Wang, X.; Wang, C.; et al., Translators; China Meteorological Press: Beijing, China, 2020; Volume 27.
28. Cao, Q.; Yu, D.Y.; Sun, Y. Research progress on the quantitative relationship between land use/cover change and climate change. *J. Nat. Resour.* **2015**, *30*, 880–890. (In Chinese)
29. Wang, J.; Qian, Y.G.; Han, L.J. Relationship between land surface temperature and land cover types based on GWR model: A case of Beijing-Tianjin-Tangshan urban agglomeration, China. *Chin. J. Appl. Ecol.* **2016**, *27*, 2128–2136. (In Chinese)
30. Buyantuyev, A.; Wu, J. Urban heat islands and landscape heterogeneity: Linking spatiotemporal variations in surface temperatures to land-cover and socioeconomic patterns. *Landsc. Ecol.* **2010**, *25*, 17–33. [CrossRef]
31. Li, S.C.; Zhao, Z.Q.; Xie, M.M.; Wang, Y.G. Investigating spatial non-stationary and scale-dependent relationships between urban surface temperature and environmental factors using geographically weighted regression. *Environ. Model. Softw.* **2010**, *25*, 1789–1800. [CrossRef]

land

Article

A Comparative Analysis of Characteristics and Synoptic Backgrounds of Extreme Heat Events over Two Urban Agglomerations in Southeast China

Xiaoyan Sun [1,2], Xiaoyu Gao [2,*], Yali Luo [1,2,*], Wai-Kin Wong [3] and Haiming Xu [1]

1 Collaborative Innovation Center on Forecast and Evaluation of Meteorological Disasters, Key Laboratory of Meteorological Disasters of Ministry of Education, Nanjing University of Information Science and Technology, Nanjing 210044, China
2 State Key Laboratory of Severe Weather, Chinese Academy of Meteorological Sciences, Beijing 100081, China
3 Hong Kong Observatory, Hong Kong 999077, China
* Correspondence: gaoxy@cma.gov.cn (X.G.); ylluo@cma.gov.cn (Y.L.)

Abstract: Based on high-resolution surface observation and reanalysis data, this paper analyzes the extreme heat events (EHEs) over two densely populated urban agglomerations in southeast China, namely the Yangtze River Delta (YRD) and the Pearl River Delta (PRD), including the spatial–temporal distribution of heatwaves and warm nights and the synoptic backgrounds for regional heatwaves. The results show that the occurrence frequency of EHEs is modulated significantly by local underlying features (i.e., land–sea contrast, terrain), and the strong nocturnal urban heat island effects make warm nights much more likely to occur in cities than rural areas during heatwaves. About 80% of the YRD regional heatwaves occur from 15 July to 15 August, while a lower fraction (53%) of the PRD heatwaves is found during this mid-summer period, which partially explains the warm-season average intensity of the former being 2–3 times the latter. A persistent, profound subtropical high is the dominant synoptic system responsible for the mid-summer YRD heatwaves, which forces significant descending motion leading to long-duration sunny weather. The mid-summer PRD heatwaves involve both high-pressure systems and tropical cyclones (TCs). A TC is present to the east of the PRD region on most (about 72%) PRD heatwave days. The organized northerly winds in the planetary boundary layer in the outer circulation of the TC transport the inland warm air, which is heated by the foehn effect at the lee side of the Nanling Mountains and possibly also the surface sensible heat flux, towards the PRD region, leading to the occurrence of the extremely high temperatures.

Keywords: extreme heat event; urban heat island; subtropical high; tropical cyclone

1. Introduction

Extreme heat events (EHEs) seriously threaten human health [1], the economy [2], and the ecosystem [3]. Their frequency and duration have gone through a dramatic increase worldwide during the past 20 years [4] and are projected to further increase in the scenario of continuing global warming [5]. It is almost certain that social issues and health-related risks induced by EHEs will bring even greater challenges to governments in the near future.

Previous studies suggest that EHEs could result from the coaction of multiple physical processes at various spatiotemporal scales, including climate variability, synoptic weather systems, and local circulations associated with the topography and inhomogeneous conditions of land surface [6]. For example, El Niño could cause extensive EHEs worldwide, especially during boreal winter, while positive phases of the North Pacific index could lead to EHEs in the southeast United States and East Asia [7]. A persistent anticyclone system often acted as the main driver of EHEs over many regions [8,9]. At higher latitudes, the blocking highs obstructed the invasion of cold air, which were responsible for the Chicago heatwave in 1995 [10], the European heatwave in 2003 [11,12], and the Russian heatwave in 2010 [13]. At mid-latitudes, the stationary subtropical highs could induce strong subsidence and dry air conditions over southeast China [14] and transport hot and dry air from the inland desert areas of central Australia to the southern part of the continent [15], favoring higher temperatures and the occurrence of EHEs. Moreover, tropical cyclones (TCs) could induce sudden and dramatic temperature rise and short-term EHEs under certain circumstances [16–18], mainly due to the subsidence-induced adiabatic heating at the periphery of the TCs' spiral rainbands. The TC circulation could also cooperate with local topography, producing the foehn effect, which causes sudden downslope heating [16]. In addition, a dry land surface usually releases more sensible heat (instead of latent heat) into the atmosphere [19] and can contribute to the occurrence of extremely high temperatures [20–22]. Substantial changes in the heat capacity and moisture of land surface due to urbanization lead to the urban heat island (UHI) effects, which could have a profound impact on EHEs over cities [23].

Over the past three decades, China has undergone rapid economic development, leading to dramatic growth in urban areas and population, especially in southeast China [24]. Two urban agglomerations, the YRD and the Pearl River Delta (PRD), are the most densely populated and flourishing economic areas in East Asia (see Figure 1 for their locations). The YRD, including the megacities of Shanghai, Nanjing, Suzhou, and Hangzhou, is an important intersection of the Belt and Road and the Yangtze River economic belt. The PRD, including the megacities of Guangzhou, Hong Kong, and Macao, is ranked as one of the four great bays in the world with diversified industries and a prosperous economy. As shown in Figure 1b,c, the underlying surface over PRD and YRD exhibits complex features with different shapes of urban agglomerations adjacent to mountains and seas. With low latitudes and wet climates, YRD and PRD experience high temperatures in boreal summer that are exacerbated by the UHI effect [25]. The YRD summer-time warming from 1979 to 2008 primarily results from the significant increase in maximum temperature, with the UHI effect accounting for 36–68% of the total regional warming [26]. Urbanization contributes to more than one-third of the increase in the intensity of EHEs in the YRD region from 1971 to 2013, which is comparable to the contribution of greenhouse gases [27]. A case study indicates that the maximum UHI intensity was up to 2.2 °C in Suzhou city during a hot weather episode from 25 July to 1 August 2007, which is much greater than the 37-year (from 1970 to 2006) summer average of 0.35 °C [28]. Considering the numerous people and infrastructures influenced by extreme temperatures, a better understanding of the characteristics and driving mechanisms of EHEs over the two regions would help in the warning and mitigation of the adverse effects of EHEs.

Numerous studies have investigated the factors and mechanisms governing the occurrence of EHEs over eastern China. Ding et al. [29] found that the occurrence frequency of heatwaves in most regions of China increased during 1961–2007, which was closely related to the changes in rainy days and atmospheric circulation patterns. Chen et al. [30] found that a strong northward water vapor flux reduced the nocturnal longwave radiation cooling around Beijing and was associated with warm nights in Beijing during July and August. Chen et al. [31] found that the warming of the western tropical Pacific forced an abnormally intensified subtropical high, which was responsible for the 2017 extremely hot mid-summer in Central and South China. Zhong et al. [32] found that frequent TC activities could indirectly lead to more hot days along the lower and middle reaches of the Yangtze River in East China, regardless of the cooling effects directly induced by TCs. Li et al. [33] showed that a persistent anticyclonic anomaly with a quasi-barotropic structure is a prerequisite for the compound heat extremes (i.e., the continuous occurrences of hot days and warm nights) in the middle and lower reaches of the Yangtze River. While the middle and lower reaches of the Yangtze River experience the highest frequency of EHEs in eastern China [29], only limited studies have focused on EHEs over the urban agglomerations, the PRD region in South China in particular.

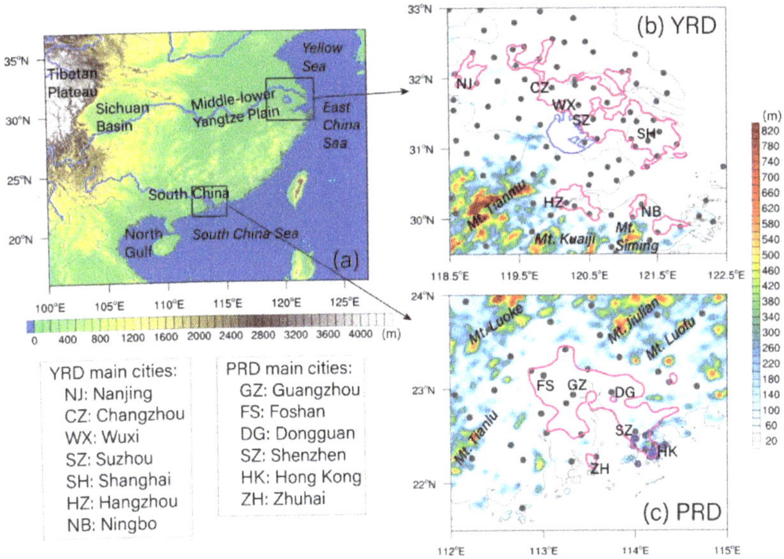

Figure 1. Topography (color shadings) over (**a**) southern China and surrounding areas and the (**b**) YRD and (**c**) PRD regions, respectively. In (**b**,**c**), gray dots denote the surface weather stations with one blue dot denoting the Hong Kong Observatory (HKO) station; pink lines denote the urban regions; names of major cities are given by abbreviation; major mountains are labeled. Blue lines in (**b**) indicate Lake Taihu. Gray lines indicate the national and provincial boundaries and coastal lines.

To fill this gap of knowledge, this study aims to analyze and compare the spatial–temporal distribution and synoptic backgrounds of EHEs over the YRD and PRD regions. Our analysis focuses on the warm seasons (May to August) in a ten-year period (2008–2017) with well-established urban agglomerations in both regions, as the rapid urbanization over the two regions started in the early to middle 1990s. The remainder of the paper is organized as follows: data and methods are introduced in Section 2; spatial–temporal distribution and synoptic backgrounds of EHEs are discussed in Sections 3 and 4, respectively; and a summary and conclusions are provided in Section 5.

2. Data and Methods

2.1. Data

The daily maximum and minimum 2 m air temperature (denoted as T_{max} and T_{min}) observed at the surface weather stations (dots in Figure 1b,c) is used to identify the EHEs. There are 93 and 46 stations in the YRD and PRD regions, respectively, including 14 stations in Hong Kong. This dataset has been quality controlled by the National Meteorological Information Center (NMIC) of China and widely used in climate studies [25,34–36]. The 5th generation of ECMWF Reanalysis (ERA5; available every hour on a $0.25° \times 0.25°$ grid and 37 vertical levels) are used to analyze the synoptic backgrounds. The TC best track dataset [37,38] provided by the China Meteorological Administration (CMA) is used to identify TCs related to the heatwaves. Moreover, the urban and rural regions are distinguished based on the fourth version of composite satellite-based nighttime lights data gathered in 2013 derived from the Defense Meteorological Satellite Program's Operational Linescan System (DMSP/OLS) of the United States (https://ngdc.noaa.gov/eog/dmsp/downloadV4composites.html, accessed on 10 June 2022). The digital numbers (DN) value of 57 is used as the urban threshold based on Wang et al. [39].

2.2. Definition and Analysis of EHEs

This study focuses on two types of EHEs, namely, heatwaves and their accompanying warm nights. According to the Glossary of Meteorology of the American Meteorological Society (AMS) [40], a heatwave is defined as a period of abnormally and uncomfortably hot and usually humid weather. The identification of heatwaves in literature involves T_{max} above a threshold during continuous days. For example, Schoetter et al. [41] focused on periods of over three consecutive days above the 98th percentile of T_{max}. The indices used to characterize heatwaves vary among different studies [10,12,41,42], which makes it difficult to quantitatively compare their results. In this study, we use a unified metric for both the YRD and PRD regions. Specifically, a period of over three consecutive days with $T_{max} > 35$ °C is identified as a heatwave, following CMA's stipulation. This definition of heatwaves is similar to that of the WMO with a lower fixed threshold of 32 °C [40]. For each weather station, we use two indices to discuss the characteristics of heatwaves, i.e., the occurrence frequency and the average cumulative intensity. The occurrence frequency of heatwave at a station is the total number of heatwaves that occurred at the station during the 10 warm seasons. The cumulative intensity of each heatwave at a station is the sum of T_{max} beyond 35 °C (i.e., T_{max}—35 °C) on each day during the heatwave, which is averaged among all the heatwaves to get the average cumulative intensity at the station. Moreover, we define a regional heatwave as a period with at least half of the stations in the region (YRD or PRD) simultaneously experiencing a heatwave. Note that, among the 14 stations in Hong Kong, only the Hong Kong Observatory station is used to analyze the regional heatwaves, because the stations in Hong Kong are distributed much more densely than other stations in the PRD (Figure 1c).

The possible occurrences of warm nights in heatwaves can aggravate the adverse impacts of heatwaves due to the prolonged high temperature in the day and night [43]. Such compound hot extremes with continuous occurrences of hot days and warm nights are recently noticed to be more frequent and intense over cities than surrounding rural areas, due to the UHI effect [44]. To define the compound EHEs, a metric is needed for warm nights. As the threshold used to define EHEs (35 °C) is about the 90th percentile of T_{max} in both the YRD and PRD, we similarly use the 90th percentile of T_{min} (approximately 27 °C) to define a warm night, i.e., a night with $T_{min} > 27$ °C. For the compound EHEs, i.e., heatwaves with warm nights, the proportion and the cumulative intensity of warm nights are calculated. The proportion is the ratio of total number of warm nights to the total number of days in a heatwave event, and the cumulative intensity of warm nights is the sum of T_{min}—27 °C during a heatwave event.

3. Characteristics of EHEs over the Two Regions

3.1. Spatial Distribution of EHEs

Substantial spatial variations of EHEs are observed over the PRD and YRD regions. In the YRD region, the occurrence frequency of heatwaves largely decreases from the mountains in the southwest to the city belt, and then to its northeast (Figure 2a). More than 30 heatwaves are observed at most mountain stations and the cities of Hangzhou and Ningbo adjacent to the mountains. The number of heatwaves reduces to 20–30 at most stations in the belt-shaped urban area. The 37 urban stations in the YRD region experience an average of 23.6 heatwaves in the 10 warm seasons, and the highest frequency of 42 is recorded in the capital city of Hangzhou. At the stations north of the city belt, the heatwave frequencies are lower than 20, likely due to the cold advection of sea breeze over the open flat land. Although much fewer heatwaves occur in the city belt than in the mountains, their intensities are roughly comparable between the two areas, mostly with a cumulative intensity of >13 °C (Figure 2b).

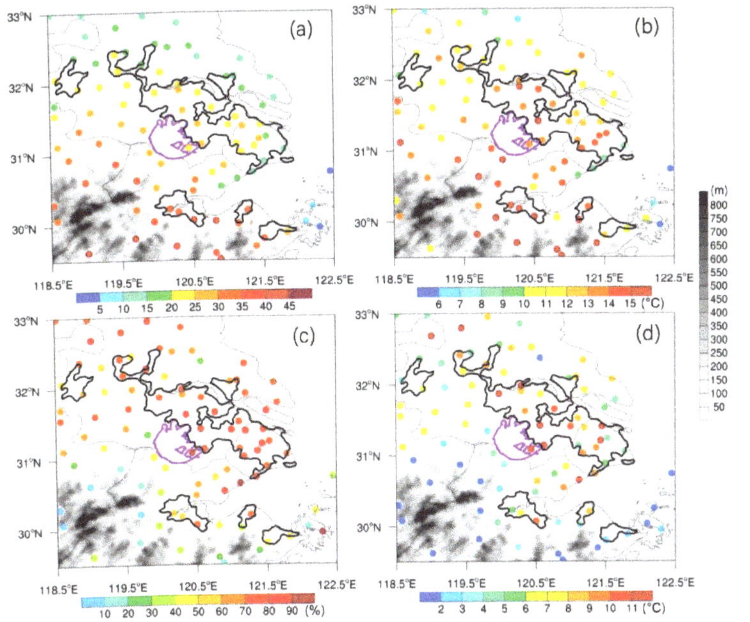

Figure 2. Spatial distribution of EHEs occurred over the YRD region in the warm seasons during 2008–2017. (**a**) Occurrence frequency of heatwaves. (**b**) Average cumulative intensity of heatwaves. (**c**) Proportion of warm nights in heatwaves. (**d**) Average cumulative intensity of warm nights. Gray shadings denote the topography. Black lines indicate the urban regions, and thin gray lines denote the provincial boundaries and coastlines. Purple lines show Lake Taihu.

The spatial distributions of warm nights during heatwaves in the YRD region exhibit a pattern that highlights the flat lands, especially the city belt (Figure 2c,d). For most urban stations, more than 70% of heatwave days are accompanied by warm nights, while this proportion is less than 40% in the mountains (Figure 2c). Meanwhile, the intensity of warm nights in the cities is much higher than that in the mountains (mostly >8 °C vs. <3 °C; Figure 2d). The proportion and cumulative intensity of warm nights averaged over the YRD urban stations are about 73% and 7.8 °C, while only 53% and 4.8 °C for the rural stations.

The heatwaves over the PRD region exhibit a clear contrast between the coastal and inland areas (Figure 3a,b). The heatwaves rarely occur (mostly <15) in the coastal area (i.e., to the south of the red line in Figure 3a, which is oriented roughly in parallel to the coastline), but sharply increase to >30 in the northern inland area. The heatwaves in the coastal area are also relatively weaker than the inland ones (the average cumulative intensity of mostly <3 °C vs. >4.5 °C). Compared to the YRD region, the intensity of heatwaves over PRD is much lower (c.f. Figures 2b and 3b), with the highest average cumulative intensity in Guangzhou city of only 5–6 °C. This is closely related to the earlier occurrences of heatwaves in June over PRD (Section 3.2) and the location of Western North Pacific Subtropical High (WNPSH) in summer (Section 4).

Similar to the YRD region, the cities in the PRD (including the coastal cities such as Hong Kong and Zhuhai) experience much larger fractions of warm nights in heatwaves than the rural areas (Figure 3c,d). The average proportion and cumulative intensity are 66% and 3.5 °C for the urban stations, while only 31% and 1.0 °C for the rural stations. These results indicate that most heatwaves in the rural areas turn into cool nights after sunset, while the nocturnal temperature remains high in the cities. These results are qualitatively consistent with the previous finding of significantly larger increasing trends of compound hot extremes than independent hot days or independent warm nights in cities over southeast China during the rapid urbanization period from the mid-1990s to 2020 [44]. The building materials of urban canopies have larger thermal conductivity and heat capacity, leading to a larger heat storage than rural areas. This slows down the nocturnal cooling and makes significant UHI effects at night [45].

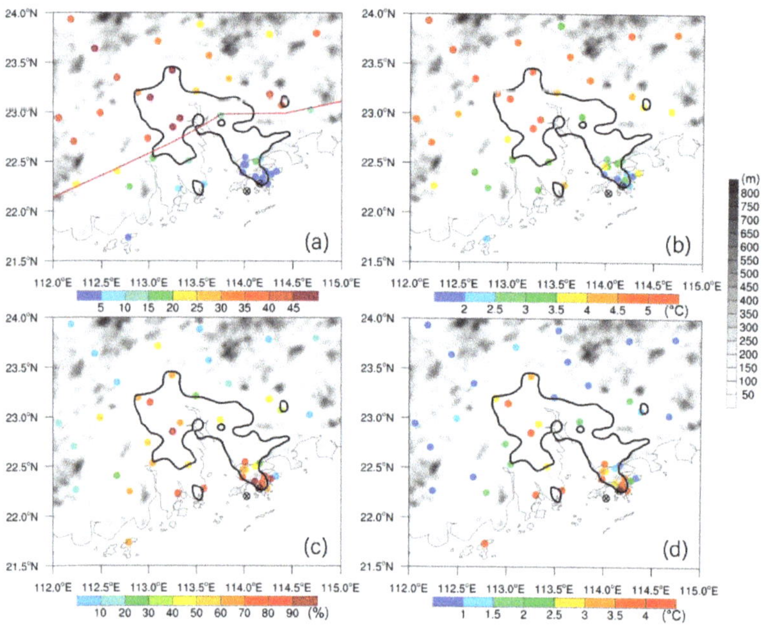

Figure 3. Spatial distribution of EHEs occurred over the PRD region in the warm seasons during 2008–2017. (**a**) Occurrence frequency of heatwaves. (**b**) Average cumulative intensity of heatwaves. (**c**) Proportion of warm nights in heatwaves. (**d**) Average cumulative intensity of warm nights. Gray shadings denote the topography. Black lines indicate the urban regions, and thin gray lines denote the provincial boundaries and coastlines. Crossed circles indicate that no heatwave occurred. The red line in (**a**) roughly separates the coastal and inland areas.

In summary, local geographical features substantially influence the occurrences of heatwaves and the impact of the UHI effect on EHEs is modulated by the presence of water and mountains over the PRD and YRD regions. The cities adjacent to inland mountains (e.g., Hangzhou in the YRD and Guangzhou in the PRD) experience the most frequent and intense heatwaves, while those adjacent to the open seas (e.g., Hong Kong and eastern Shanghai) show less frequent heatwaves with weaker intensity (Figures 2a and 3a). The urban and rural stations sharing similar geography observe smaller differences in the heatwave frequency (Figures 2a and 3a), but both inland and coastal cities experience much more warm nights in heatwaves (Figures 2c and 3c) with substantially higher intensity (Figures 2d and 3d) than the rural areas. The result emphasizes the important role of the UHI effect in producing nocturnal high temperatures in both regions. Moreover, the intensities of heatwaves and the accompanied warm nights over the YRD are about 2–3 times higher than those in the PRD (c.f., Figures 2b,d and 3b,d), which can be attributed to the differences in the dates and synoptic backgrounds of the heatwaves between the two regions to be shown in the following sections.

3.2. Temporal Variation of the Regional Heatwaves

In total, 137 days of the YRD regional heatwave and 102 days of the PRD regional heatwave are found during the 10 warm seasons, and all of them occur in summer (JJA).

The YRD regional heatwaves begin on 1 July and end on 23 August, and the PRD regional heatwaves begin on 1 June and end on 29 August (Figure 4a). It follows that occurrences of the heatwaves in the YRD region are more concentrated in mid-summer, while those in the PRD occur also in June and near the end of August. About 80% of the YRD regional heatwave days and 53% of the PRD regional heatwave days are from 15 July to 15 August (referred to as 'mid-summer' hereafter). Only 13.1% and 7.3% of the YRD regional heatwave days occur before and after the mid-summer, respectively, while the proportions in the PRD are nearly doubled or more (33.3% and 13.7%). The PRD is located in lower latitudes than the YRD, which could largely explain the earlier and later occurrences of heatwaves in the PRD. In June, South China is at the late stage of its early-summer rainy season [46] and under the influence of the WNPSH, while the Yangtze River basin is out of the reach of WNPSH. Thus, heatwaves could strike the PRD but hardly influence YRD in June. In mid-summer, the WNPSH often covers the middle and lower reaches of the Yangtze River, favoring the formation of the YRD heatwaves, while the PRD is to the southwest of WNPSH and tropical systems occasionally influence southeast China. These indicate that the relevant synoptic backgrounds of heatwaves over the YRD and PRD in mid-summer might be different. The following section will focus on synoptic patterns governing the YRD and PRD regional heatwaves during mid-summer.

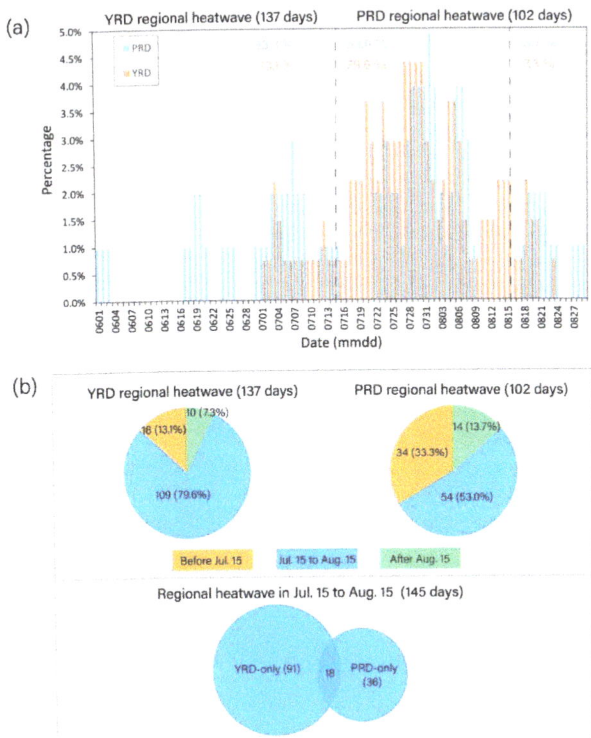

Figure 4. Occurrence frequency of regional heatwaves over the PRD and YRD regions, respectively, during the 10 warm seasons. (**a**) Fraction of heatwaves on each day to the total heatwave days during the warm season. (**b**) Numbers (fractions; unit: %) of the regional heatwave days in the entire warm season and a variety of subperiods, respectively.

4. Synoptic Backgrounds for the Mid-Summer YRD- and PRD-Regional Heatwaves

In mid-summer, 109 and 54 regional heatwave days are found over the YRD and PRD regions, respectively, including 91 days of the YRD-only heatwave, 36 days of the PRD-only heatwave, and 18 days of simultaneous heatwaves over both regions. Synoptic patterns are composited for each of the three groups and their subgroups. Specifically, the geopotential height (GPH), total cloud cover, net solar radiation at the surface, horizontal wind vector, and vertical wind velocity available hourly in the ERA5 dataset are averaged to produce their daily values. Then, for each group and/or subgroup of the regional heatwaves, all the daily values are averaged to show their composite synoptic backgrounds. The anomalies are defined as the differences between the heatwave days and mid-summer. Anomalies of geopotential height shown in the following figures have passed the Student t-test of 95% significance.

4.1. The YRD-Only Heatwaves

At 300 hPa, the YRD region is situated in an extensive, east–west-oriented, high-pressure area with GPH > 9740 gpm, downward motion, and low horizontal wind speeds (Figure 5a). The positive anomaly of geopotential height over and to the north of the YRD is accompanied by easterly anomalies (Figure 5d), reflecting that the upper-level westerly jet in mid-summer is weakened during the YRD regional heatwaves.

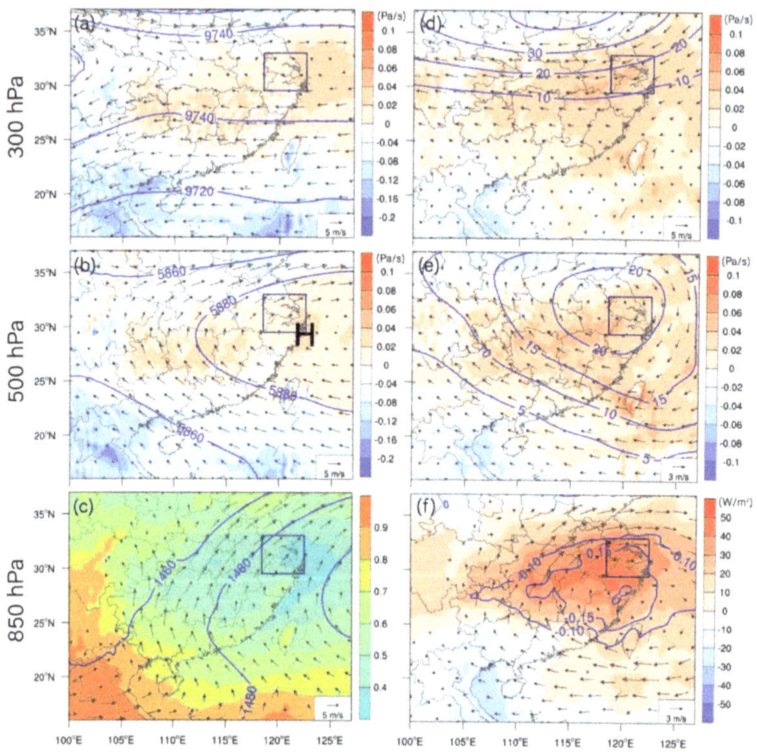

Figure 5. Synoptic backgrounds for the YRD-only heatwaves. (**a**) The 300 hPa geopotential height (blue contours; unit: gpm), horizontal wind vector (arrows), and vertical velocity (shadings). (**b**) Same as (**a**), but at 500 hPa. (**c**) Geopotential height (blue contours; unit: gpm) and horizontal wind vector (arrows) at 850 hPa, and total cloud cover (shadings). (**d**) Anomalies of geopotential height (blue contours; unit: gpm), horizontal wind vector (arrows), and vertical velocity (shadings) at 300 hPa. (**e**) Same as (**d**) but at 500 hPa. (**f**) Anomalies of total cloud cover (blue contours), 850 hPa horizontal wind vectors (arrows), and net solar radiation at the surface (shadings; positive downward). Navy rectangle denotes the YRD region. Gray lines indicate the national and provincial boundaries and coastal lines.

In the middle troposphere (500 hPa), the strong WNPSH extends westward to cover the middle and lower reaches of the Yangtze River (Figure 5b), accompanied by descending motion over the region (Figure 5a,b) with large anomalies of both GPH and vertical velocity (Figure 5d,e).

The lower troposphere (850 hPa) also experiences a strong high pressure extending from the North Pacific with GPH > 1480 gpm over the YRD (Figure 5c). Associated with the easterly wind anomalies over the northern South China Sea (Figure 5e,f), water vapor transport toward the YRD region is weakened relative to the strong southwesterly monsoonal flow usually prevailing over southeastern China in mid-summer. The drier lower and middle troposphere accompanied by the descending motion leads to smaller cloud cover (<0.4; Figure 5c), which is 0.15 less than the mid-summer average, and consequently positive anomalies of the net short-wave radiation at the surface over the YRD region (>40 W m^{-2}; Figure 5f).

In short, during the YRD-only heatwaves, the upper-troposphere jet stream moves northward to the north of 35° N with low wind speeds over the YRD; the WNPSH in the middle and lower troposphere extends westward to the middle and lower reaches of the Yangtze River, leading to descending motion and drier atmosphere over the YRD region. As a result, much fewer clouds are formed and more solar radiation is absorbed at the land surface. The increased solar heating and descending-induced adiabatic heating jointly lead to the high temperature and heatwaves over the YRD region. These results are consistent with previous studies on regional compound hot extremes over the middle and lower reaches of the Yangtze River [33].

4.2. The PRD-Only Heatwaves

Compared to the above results, synoptic patterns during the PRD-only regional heatwaves (36 days) differ substantially. At 300 hPa, the GPH over the East China Sea and Taiwan Island is significantly lower than the mid-summer average (Figure 6d), although a closed low-pressure center is not evident in the composite field (Figure 6a). In associated with the negative anomalies of GPH, northwesterly anomalies are observed over the PRD region, which leads to weakening of the tropical easterly jet stream that is often present over the PRD in mid-summer (not shown) and the prevailing northeasterly winds instead (Figure 6a).

The composite fields in the middle and lower troposphere (Figure 6b,c) illustrate a cyclonic circulation centered around Taiwan Island. The low-pressure system centered around Taiwan Island (Figure 6b) is accompanied by negative anomalies of the 500 hPa GPH (up to 30–40 gpm lower than the mid-summer average; Figure 6e) and cyclonic anomalies in the middle and lower troposphere (Figure 6e,f). The prevailing low-level southwesterly winds over the PRD in mid-summer are substantially weakened during the PRD-only heatwaves (Figure 6b,c). The northerly anomalies during the PRD-only regional heatwaves significantly decrease the moisture needed for the production of clouds and precipitation, as the low-level southwesterly wind from the South China Sea is the key provider of water vapor over the PRD region.

The presence of cyclonic circulations in the middle to lower troposphere (Figure 6b,c) suggests a possible impact of TCs on the PRD-only regional heatwaves. A closer examination indicates that TCs are present over the offshore waters and/or the coastal areas of southeast China during 81% (29 out of 36) of the PRD-only heatwave days. The ten TCs (Table 1) mostly move northwestward from the western North Pacific and make landfalling in Fujian Province after passing the Taiwan strait except for Typhoon Nida (1604), Muifa (1109), and Nakri (1412) (Figure 7). Typhoon Nida moves towards the PRD region from the South China Sea, while Typhoon Muifa and Typhoon Nakri move northward to the Yellow Sea without landfalling during the PRD heatwave days. Considering the changes in the synoptic conditions with different locations of the TCs, composite analyses are conducted for four subgroups, respectively, i.e., the 12, 11, 2, and 4 days of the PRD-only regional heatwave accompanied by TCs centered in the four regions (A to D in Figure 7). Note that the TCs in Region A or B move into Region C or D later, and the TCs centered in Regions A and B are relatively stronger (Figure 7). Except for the four subgroups of the PRD-only heatwave with TCs, the other 7 days do not experience TCs, and the synoptic patterns on these days will be discussed as the fifth subgroup.

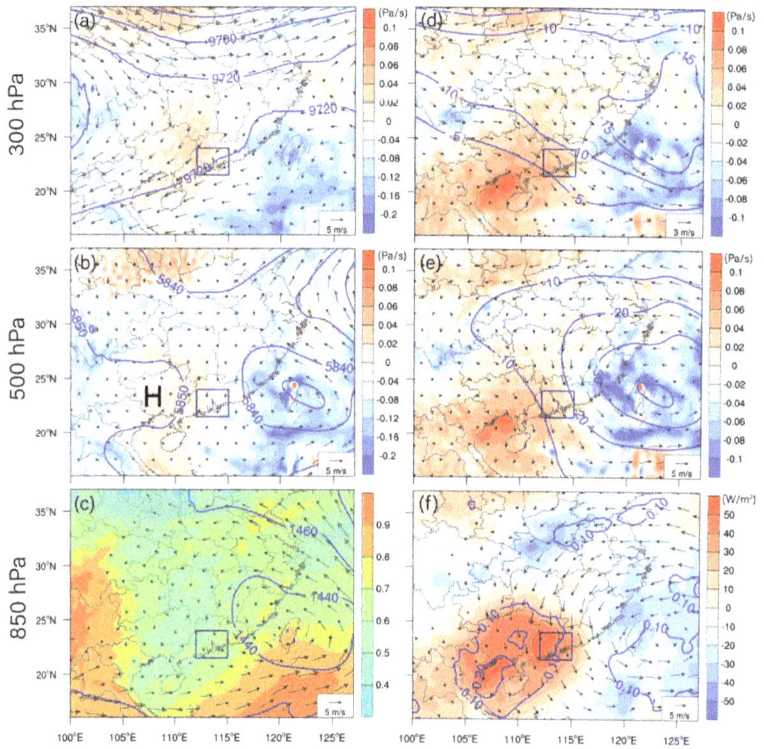

Figure 6. Synoptic backgrounds for the PRD-only heatwaves. (**a**) The 300 hPa geopotential height (blue contours; unit: gpm), horizontal wind vector (arrows), and vertical velocity (shadings). (**b**) Same as (**a**), but at 500 hPa. (**c**) Geopotential height (blue contours; unit: gpm) and horizontal wind vector (arrows) at 850 hPa, and total cloud cover (shadings). (**d**) Anomalies of geopotential height (blue contours; unit: gpm), horizontal wind vector (arrows), and vertical velocity (shadings) at 300 hPa. (**e**) Same as (**d**) but at 500 hPa. (**f**) Anomalies of total cloud cover (blue contours), 850 hPa horizontal wind vectors (arrows), and net solar radiation at the surface (shadings; positive downward). Navy rectangle denotes the PRD region. Gray lines indicate the national and provincial boundaries and coastal lines.

The synoptic backgrounds for the five subgroups of the PRD-only heatwave days are shown in Figure 8. With the TCs centered in Region A, the PRD region is situated about 500 km west of the TC center with low-level northerly winds aloft. Note that with the TCs in their mature stage in Region A, the strong upward motion in the cyclonic circulations could force extensively distributed descending to its northwest over the drier inland areas. The PRD region Is located between the ascending and descending regions, where the air vertical velocity seems negligible. Thus, unlike the YRD regional heatwaves induced by prominent high-pressure systems, the synoptic subsidence-induced adiabatic heating is not very strong over the PRD region and the cloud cover over the PRD region is still large (0.7–0.9), suggesting reduced solar radiative heating at the surface. Instead, the heatwaves over the PRD region are closely related to the organized low-level northerly airflows, which transport hotter air from inland mountains to the coastal area of South China (see Figure 1c for the topography) and the related foehn wind. These relevant physical processes can be seen clearly in Figure 9, which shows the vertical thermal and dynamic structure over the PRD region and the Nanling Mountains to its north during the PRD heatwave days with the TCs in Region A. The near-surface air near the PRD coastline is about 3 K cooler than the northern PRD region, likely due to the large tropical land–sea thermal contrast in mid-summer. Meanwhile, the TCs to the east force northerly winds over South China. When the northerly airflows climb over the Nanling Mountains, the adiabatic cooling leads to the formation of precipitating clouds with liquid water content (q_c) of >0.04 g kg^{-1} over the mountains. At the lee side, descending compression causes strong adiabatic heating, as much water vapor has already condensed and been taken away by rainfall. This TC-related foehn effect [16,47,48] increases air temperature at the lee side of the Nanling Mountains, making the northern boundary of the PRD region (at about 24° N) a warm center. This thermal structure combined with the prevailing northerly winds closely associated with the TCs produces intense warm advection over the PRD region, favoring the occurrence of the PRD regional heatwaves. Moreover, the upstream surface sensible heat flux and TC-induced subsidence in its outer circulation could also be contributing factors, as previously documented in a case study [48].

Table 1. Basic information about the TCs on the PRD regional heatwave days.

TC Number	TC's Name	TC's Life Span (mmddhh UTC)
0808	Fung-wong	072400-073118
0908	Morakot	080306-081306
1109	Muifa	072700-080900
1209	Saola	072618-080500
1410	Matmo	071700-072600
1412	Nakri	072900-080400
1513	Soudelor	073000-081206
1604	Nida	072912-080300
1709	Nesat	072500-073100
1710	Haitang	072712-080306

When the TCs move from Region A into Region C, the PRD region is located to their south with westerly to southwesterly airflows and weak ascent in the middle and lower troposphere (Figure 8c). The two days of such PRD-only regional heatwave are the last stage of heatwaves induced by TCs in Region A. When the weakened TCs move further northward, the direction of isobaric lines over the PRD region change back to southwest-northeast, and the cooler air brought by the southwesterly winds from the South China Sea rapidly ends the heatwaves.

Compared to the TCs in Region A, the TCs centered in Region B possess similar intensity, but are located several hundred kilometers to the east with the cyclonic circulation hardly extending to the west of Taiwan Island (Figure 8b). Meanwhile, an extensive high-pressure region extends from southeast China to the Yellow Sea (Fee Figure 1a for its location), leading to descending motion and sunny weather. The strong TCs in Region B could contribute to the persistence of the high-pressure system by causing compensating subsidence. Situated roughly at the high-pressure center, the PRD region suffers from heatwaves due to the synoptic subsidence-induced adiabatic heating and enhanced clear-sky solar heating at the surface. The high-pressure system lasts a few days and vanishes after the TCs move from Region B into Region D (Figure 8d). Then, the descending motion over the PRD region is also weakened, while the southwesterly winds over the South China Sea are substantially strengthened (c.f. Figure 8b,d). Consequently, the heatwave over the PRD region is about to end.

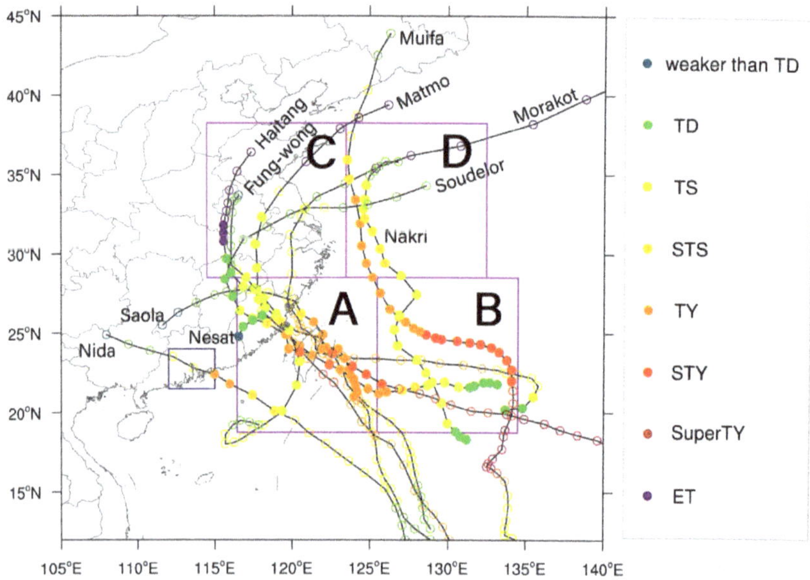

Figure 7. Tracks of the ten TCs associated with the PRD-only regional heatwave. The filled and hollow circles denote the TC centers at 6 h intervals on the heatwave days and the other days, respectively, with colors denoting the intensity of TC estimated by CMA: tropical depression (TD), tropical storm (TS), severe tropical storm (STS), typhoon (TY), severe typhoon (STY), super typhoon (SuperTY), and extratropical transition (ET). Purple rectangles denote the four regions where the TCs are centered. Navy rectangle shows the PRD region. Gray lines indicate the national and provincial boundaries and coastlines. A to D represent four regions divided according to different locations of TCs.

The PRD heatwaves without TCs in the western North Pacific is usually associated with a subtropical high over coastal areas of South China and the adjacent water surface (Figure 8e), which explains the descending motion and anomalies over the North Gulf in Figure 6b,e. This subtropical high seems smaller in coverage and weaker in intensity than the strong WNPSH responsible for the YRD heatwaves (Figure 5b), but is comparable to the high-pressure system on the PRD heatwave days with TCs in the Region B (Figure 8b). Under the control of this subtropical high and associated downward motion, heatwaves occur over the PRD region.

4.3. The YRD and PRD Simultaneous Heatwaves

Synoptic analysis for the 18 days with simultaneous heatwaves over both the YRD and PRD regions suggests a combination of the major features observed during the PRD-only heatwaves and the YRD-only heatwaves, respectively (Figure 10). Synoptic fields are composited for the 10 days with concurrent TCs and the other 8 days without TC, respectively (Figure 10). As shown in Figure 10a–c, while the WNPSH extends westward to cover the YRD region, the PRD region is situated in the outer circulation of TCs centered several thousand kilometers to the southeast. The dominance of WNPSH over the YRD region is to some extent similar to that during the YRD-only heatwaves (Figure 5a–c) with about the same magnitudes of descent and even smaller cloud cover. On the other hand, the impacts of TCs to the PRD region resemble what occurs on the PRD-only heatwave days with TCs over Region A (Figure 8a) despite that the TCs are relatively displaced southward in the middle troposphere due to the westward extending WNPSH over the YRD (Figure 10b), i.e., the TC-induced foehn effect associated with Nanling Mountains, and possibly the upstream surface sensible heat flux and subsidence induced by the TCs probably play key roles. Note that about 72% (39 out of 54) of the mid-summer PRD regional heatwave days are accompanied by TCs.

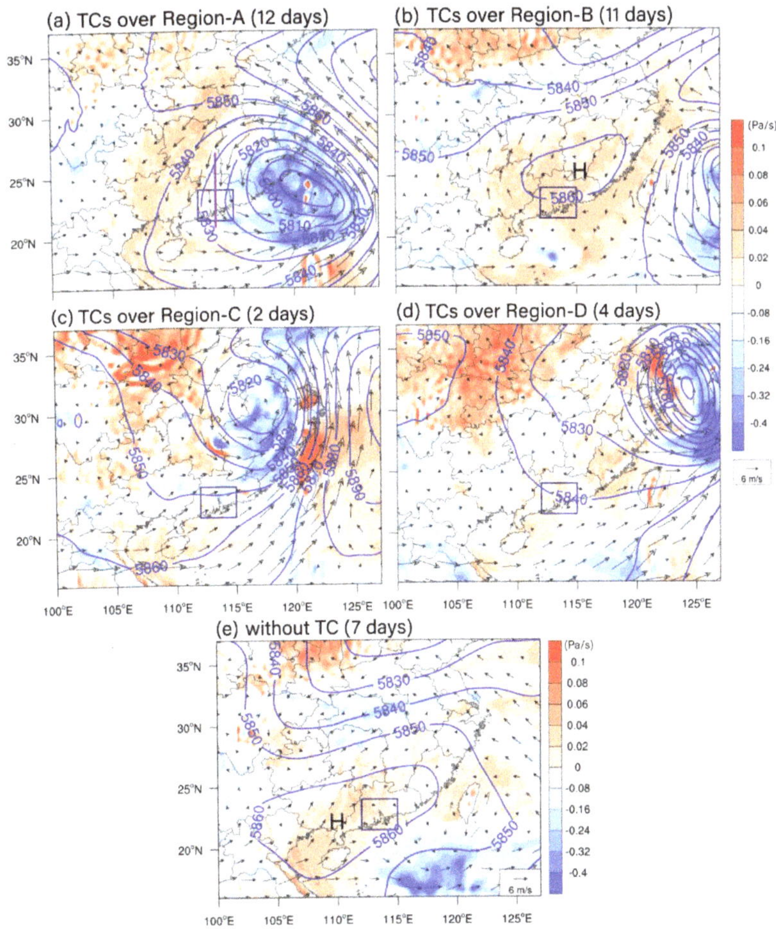

Figure 8. Synthetic fields for the five subgroups of the PRD-only heatwave days: (**a**–**e**) with TCs centered over Regions A, B, C, and D as shown in Figure 7 and I without TC, including 500 hPa geopotential height (blue contours; unit: gpm) and vertical velocity (shadings), and 850 hPa horizontal wind vectors (arrows). Thick purple line in (**a**) represents the location of vertical cross-section used in Figure 9. Rectangular box in each panel denotes the PRD region. Gray lines show the national and provincial boundaries and coastlines.

The other 8 days are characterized by an extremely intense, extensive subtropical high throughout the troposphere, leading to downward motion and few clouds over most areas of southeast China (Figure 10d–f). At 500 hPa, this high-pressure system is the intensified WNPSH and centered roughly between YRD and PRD (Figure 10e). These synoptic features on the 8 days (including 6 days in late July 2016 and 2 days in early August 2017) are like a magnified version of those appearing during the YRD-only heatwave days (Figure 5a–c). They could be related to the weakening of the Somalia cross-equatorial flows and the westerly flow over the tropical Indian Ocean [49].

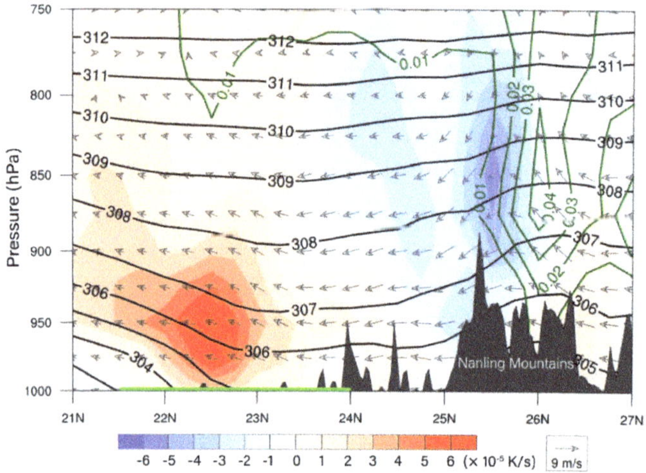

Figure 9. Vertical cross-section along the thick purple line in Figure 8a for the PRD-only heatwave days with a TC centered in Region A. Arrows denote the wind vectors along the plane. The vertical velocity is multiplied by a factor of 200. Black contours denote the potential temperature (θ; unit: K). Green contours denote the mixing ratio of cloud liquid water (q_c; unit: g kg^{-1}). Shadings denote the horizontal advection of θ. Gray shading denotes topography. Thick green line indicates the location of the PRD region.

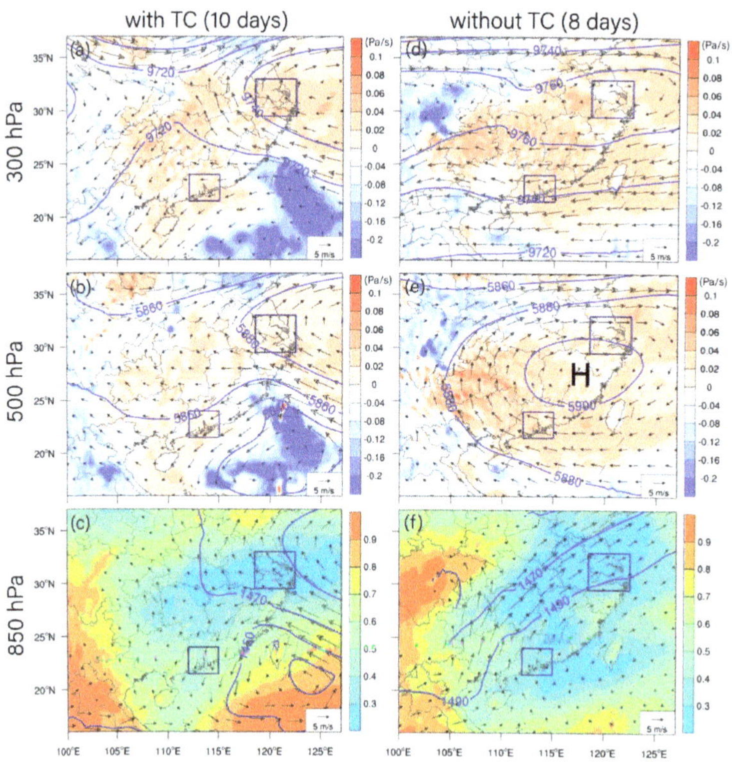

Figure 10. Synoptic backgrounds for the heatwaves simultaneously over both YRD and PRD regions

(rectangular boxes), with TC (left) and without TC (right), respectively. Top panels (**a**,**d**) represent the 300 hPa GPT (blue contours; unit: gpm), horizontal wind vector (arrows), and vertical velocity (shadings; positive and negative value corresponds to descending and ascending motion). Middle panels (**b**,**e**) are same as the top panels except for at 500 hPa. Bottom panels (**c**,**f**) represent the 850 hPa GPH (blue contours; unit: gpm) and horizontal wind vector (arrows), and total cloud cover (shadings). Gray lines show the national and provincial boundaries and coastlines.

5. Summary and Conclusions

EHEs exert adverse effects on human health, the economy, and the ecosystem, and have increased with climate warming and global urbanization. Given the large population exposed to the high temperatures in big cities, it is important to deepen the understanding of the characteristics and mechanisms of urban EHEs.

The YRD and PRD in the coastal areas of southeast China are the most flourishing economic areas in East Asia that suffer from EHEs in boreal summer. Using the observation of dense surface weather stations and high-resolution reanalysis data, this study comparatively analyzes EHEs over the two urban agglomerations in ten warm seasons (May to August 2008–2017). Unified metrics are used to define and characterize heatwaves and warm nights at the meteorological stations in both regions, and their spatial–temporal distributions are revealed. Moreover, the regional heatwave of each region is defined based on the proportion (>0.5) of stations simultaneously going through a heatwave in the region. Synoptic backgrounds for the regional heatwaves over the YRD only, PRD only, and both the YRD and PRD simultaneously are discussed in detail. The major conclusions are as follows.

(1) Both regions observe substantial spatial variations of heatwaves, reflecting the modulation of the local geographic environment on the characteristics of heatwaves. The inland mountains experience the most frequent heatwaves, while the coasts experience the least due to the cold advection of sea breeze. Despite little difference in heatwave frequency between urban and rural areas with similar geographic environments, cities suffer from much more frequent and intense warm nights in heatwaves than the rural areas in both the YRD and PRD.

(2) All the regional heatwaves over the YRD and PRD regions occur in summer (JJA). Those over the YRD are more concentrated in mid-summer (from 15 July to 15 August), while those over the PRD start earlier in June and end later at the end of August. This partially explains the more intense heatwaves in the YRD region with a warm-season average cumulative intensity about 3 times that in the PRD region.

(3) The persistent, enhanced WNPSH in the middle to lower troposphere is the dominant system responsible for the mid-summer YRD heatwaves, which account for about 80% of the YRD heatwaves in the ten warm seasons. The strong WNPSH extends over the mid-lower reaches of the Yangtze River, forcing a significant descending motion. Meanwhile, the water vapor transport to the YRD region is weakened, as the prevailing southwesterly airflows over South China are replaced by southeasterlies. Consequently, much more solar radiation (>40 W m^{-2}) is absorbed at the land surface than the mid-summer average. The descending adiabatic heating and enhanced solar heating jointly produce heatwaves. These are generally consistent with previous studies [33,50].

(4) About 53% of the PRD heatwave days are in mid-summer and influenced by both TCs and subtropical high-pressure systems. Among those days, about 72% experience TCs to the east of the PRD region. With the TCs centered around Taiwan Island, organized low-level northerly airflows prevail over the PRD and inland areas to its north. The warm air climbs over the Nanling Mountains and is heated at the lee side by the foehn effect and likely the surface sensible heat flux, and further advected to the PRD region. On the PRD heatwave days with a remote TC to the east of Taiwan Island or without TCs, a local high-pressure system is the major driver.

(5) Regional heatwaves could occur simultaneously over both the YRD and PRD under two types of synoptic configurations. One is characterized by an extremely intense WNPSH covering southeast China; the other features a coupling between an enhanced WNPSH extending the over YRD and TCs centered to the south of Taiwan Island causing northerly low-level airflows over the PRD and the Nanling Mountains to its north.

This study clearly demonstrates that urbanization has increased the occurrence of warm nights over the two major urban agglomerations in southeast China, which could aggravate the detrimental impacts of heatwaves. This result is consistent with the previous findings of Chen and Zhai [43] and Gao et al. [44] and calls on the government, enterprises, and the public to take joint action to reduce the urban heat island effect. Possible measures include planting more green plants in urban areas, producing and using more energy-efficient air conditioners and cars, and building residential buildings with good ventilation and heat dissipation functions. Moreover, the statistical analysis in this study quantitatively reveals the contributions of both TCs and high-pressure systems to the mid-summer EHEs over the PRD, and supplements the previous finding about the role of the TC-induced foehn effect in the occurrences of EHEs in the downslope regions of Taiwan [16] and southeastern coast of China [48]. As TCs are projected to be more intense with global warming [49,50], the adverse impacts of TC-induced EHEs on global society might increase [51]. However, changes in EHEs over the PRD region and other cities worldwide in the future are unknown and deserve further study, as such regional EHEs depend on the track and movement of TCs, whose projection is associated with large uncertainties.

We should acknowledge the limitations of this study and point out that more investigations are needed in the future. The number of EHEs can be increased by extending the analysis period to improve the statistical significance and robustness of the conclusions. For the PRD heatwaves accompanied by TCs centered around Taiwan Island, the possible role played by the upstream surface sensible heat flux is speculated based on a previous study for an extreme heat event induced by Typhoon Lekima [48], and quantitative analysis for our cases is needed. Further investigation on the duration of the heatwaves under different circulation patterns, predictability of the circulation patterns, and their linkage with atmospheric variability modes and sea surface temperature is important and should help the prediction of the EHEs in the PRD and YRD regions. Furthermore, using the thermal biometeorological indices widely used in thermal discomfort assessment such as PET and UTCI [52] would supplement the characteristics of EHEs based on temperature only.

Author Contributions: Conceptualization, Y.L., X.G. and X.S.; methodology, Y.L., X.G. and X.S.; software, X.S.; validation, Y.L., X.G. and X.S.; formal analysis, X.S.; investigation, X.S. and X.G.; resources, Y.L. and X.S.; data curation, X.S.; writing—original draft preparation, X.G. and X.S.; writing—review and editing, Y.L., X.G., X.S., W.-K.W. and H.X.; visualization, X.S.; supervision, X.G. and Y.L.; project administration, Y.L.; funding acquisition, Y.L. All authors have read and agreed to the published version of the manuscript.

Funding: This research was funded by the Guangdong Major Project of Basic and Applied Basic Research (2020B0301030004) and the National Natural Science Foundation of China, grant numbers 42030610 and 41975106.

Data Availability Statement: The ERA5 Reanalysis dataset is available at https://cds.climate.copernicus.eu/cdsapp#!/home (accessed on 5 September 2022); DMSP/OLS nighttime lights data are available at https://www.noaa.gov/ (accessed on 10 June 2022); the TC best track dataset is available at https://tcdata.typhoon.org.cn/ (accessed on 30 October 2022). Temperature data were provided by the National Meteorological Information Center and the Hong Kong Observatory, which are available from the authors upon reasonable request.

Conflicts of Interest: The authors declare no conflict of interest.

References

1. Mora, C.; Dousset, B.; Caldwell, I.R.; Powell, F.E.; Geronimo, R.C.; Bielecki, C.R.; Counsell, C.W.; Dietrich, B.S.; Johnston, E.T.; Louis, L.V.; et al. Global risk of deadly heat. *Nat. Clim. Chang.* **2017**, *7*, 501–507. [CrossRef]
2. Miller, S.; Muir-Wood, R.; Boissonnade, A. An exploration of trends in normalized weather-related catastrophe losses. In *Climate Extremes and Society*; Diaz, H., Murnane, R., Eds.; Cambridge University Press: Cambridge, UK, 2008; pp. 225–247.
3. Karoly, D.J. The recent bushfires and extreme heatwave in southeast Australia. *Bull. Aust. Meteorol. Oceanogr. Soc.* **2009**, *22*, 10–13.
4. Christidis, N.; Jones, G.S.; Stott, P.A. Dramatically increasing chance of extremely hot summers since the 2003 European heat wave. *Nat. Clim. Chang.* **2015**, *5*, 46–50. [CrossRef]
5. IPCC AR6 Climate Change 2022: Impacts, Adaptation and Vulnerability. Available online: https://www.ipcc.ch/report/ar6/wg2/downloads/report/IPCC_AR6_WGII_FullReport.pdf (accessed on 7 November 2022).
6. Perkins, S.E. A review on the scientific understanding of heatwaves-Their measurement, driving mechanisms, and changes at the global scale. *Atmos. Res.* **2015**, *164*, 242–267. [CrossRef]
7. Kenyon, J.; Hegerl, G.C. Influence of modes of climate variability on global temperature extremes. *J. Clim.* **2008**, *21*, 3872–3889. [CrossRef]
8. Cowan, T.; Hegerl, G.C.; Colfescu, I.; Bollasina, M.; Purich, A.; Boschat, G. Factors contributing to record-breaking heat waves over the Great Plains during the 1930s Dust Bowl. *J. Clim.* **2017**, *30*, 2437–2461. [CrossRef]
9. Luo, M.; Lau, N. Heat waves in southern China: Synoptic behavior, long-term change, and urbanization effects. *J. Clim.* **2017**, *30*, 703–720. [CrossRef]
10. Meehl, G.A.; Tebaldi, C. More intense, more frequent, and longer lasting heat waves in the 21st century. *Science* **2004**, *305*, 994–997. [CrossRef]
11. Black, E.; Blackburn, M.; Harrison, G.; Hoskins, B.; Methven, J. Factors contributing to the summer 2003 European heatwave. *Weather* **2004**, *59*, 217–223. [CrossRef]
12. Vautard, R.; Gobiet, A.; Jacob, D.; Belda, M.; Colette, A.; Déqué, M.; Fernández, J.; García-Díez, M.; Goergen, K.; Güttler, I.; et al. The simulation of European heat waves from an ensemble of regional climate models within the EURO-CORDEX project. *Clim. Dyn.* **2013**, *41*, 2555–2575. [CrossRef]
13. Matsueda, M. Predictability of Euro-Russian blocking in summer of 2010. *Geophys. Res. Lett.* **2011**, *38*, L06801. [CrossRef]
14. Chen, R.; Wen, Z.; Lu, R. Large-scale circulation anomalies and intraseasonal oscillations associated with long-lived extreme heat events in South China. *J. Clim.* **2018**, *31*, 213–232. [CrossRef]
15. Marshall, A.G.; Hudson, D.; Wheeler, M.C.; Alves, O.; Hendon, H.H.; Pook, M.J.; Risbey, J.S. Intra-seasonal drivers of extreme heat over Australia in observations and POAMA-2. *Clim. Dyn.* **2014**, *43*, 1915–1937. [CrossRef]
16. Chen, T.C.; Wang, S.Y.; Yen, M.C.; Clark, A.J.; Tsay, J.D. Sudden surface warming-drying events caused by typhoon passages across Taiwan. *J. Appl. Meteorol. Clim.* **2010**, *49*, 234–252. [CrossRef]
17. Shibata, Y.; Kawamura, R.; Hatsushika, H. Role of large-scale circulation in Triggering Foehns in the Hokuriku District of Japan during Midsummer. *J. Meteorol. Soc. Jpn.* **2010**, *88*, 313–324. [CrossRef]
18. Parker, T.J.; Berry, G.J.; Reeder, M.J. The influence of tropical cyclones on heat waves in Southeastern Australia. *Geophys. Res. Lett.* **2013**, *40*, 6264–6270. [CrossRef]
19. Alexander, L. Extreme heat rooted in dry soils. *Nat. Geosci.* **2010**, *4*, 12–13. [CrossRef]
20. Seneviratne, S.I.; Lüthi, D.; Litschi, M.; Schär, C. Land–atmosphere coupling and climate change in Europe. *Nature* **2006**, *443*, 205–209. [CrossRef]
21. Lorenz, R.; Jaeger, E.B.; Seneviratne, S.I. Persistence of heat waves and its link to soil moisture memory. *Geophys. Res. Lett.* **2010**, *37*, L09703. [CrossRef]
22. Oke, T.R.; Johnson, G.T.; Steyn, D.G.; Watson, I.D. Simulation of surface urban heat islands under 'ideal' conditions at night Part 2: Diagnosis of causation. *Bound Lay Meteorol.* **1991**, *56*, 339–358. [CrossRef]
23. Fischer, E.M.; Oleson, K.W.; Lawrence, D.M. Contrasting urban and rural heat stress responses to climate change. *Geophys. Res. Lett.* **2012**, *39*, 2011GL05057. [CrossRef]
24. Zhao, S.; Zhou, D.; Zhu, C.; Qu, W.; Zhao, J.; Sun, Y.; Huang, D.; Wu, W.; Liu, S. Rates and patterns of urban expansion in China's 32 major cities over the past three decades. *Landsc. Ecol.* **2015**, *30*, 1541–1559. [CrossRef]
25. Liao, W.; Liu, X.; Li, D.; Luo, M.; Wang, D.; Wang, S.; Baldwin, J.; Lin, L.; Li, X.; Feng, K.; et al. Stronger contributions of urbanization to heat wave trends in wet climates. *Geophys. Res. Lett.* **2018**, *45*, 11310–11317. [CrossRef]
26. Wu, K.; Yang, X. Urbanization and heterogeneous surface warming in eastern China. *Chin. Sci. Bull.* **2013**, *58*, 1363–1373. [CrossRef]
27. Yang, X.; Ruby Leung, L.; Zhao, N.; Zhao, C.; Qian, Y.; Hu, K.; Liu, X.; Chen, B. Contribution of urbanization to the increase of extreme heat events in an urban agglomeration in east China. *Geophys. Res. Lett.* **2017**, *44*, 6940–6950. [CrossRef]
28. Zhang, N.; Zhu, L.F.; Zhu, Y. Urban heat island and boundary layer structures under hot weather synoptic conditions: A case study of Suzhou city, China. *Adv. Atmos. Sci.* **2011**, *28*, 855–865. [CrossRef]
29. Ding, T.; Qian, W.; Yan, Z. Changes in hot days and heat waves in China during 1961–2007. *Int. J. Climatol.* **2010**, *30*, 1452–1462. [CrossRef]
30. Chen, R., Lu, R. Large-scale circulation anomalies associated with 'tropical night' weather in Beijing, China. *Int. J. Climatol.* **2014**, *34*, 1980–1989. [CrossRef]

31. Chen, R.; Wen, Z.; Lu, R.; Wang, C. Causes of the extreme hot midsummer in central and South China during 2017: Role of the western tropical pacific warming. *Adv. Atmos. Sci.* **2019**, *36*, 465–478. [CrossRef]
32. Zhong, Z.; Chen, X.; Yang, X.Q.; Ha, Y.; Sun, Y. The relationship of frequent tropical cyclone activities over the western North Pacific and hot summer days in central-eastern China. *Theor. Appl. Climatol.* **2019**, *138*, 1395–1404. [CrossRef]
33. Li, Y.; Ding, Y.; Liu, Y. Mechanisms for regional compound hot extremes in the mid-lower reaches of the Yangtze River. *Int. J. Climatol.* **2020**, *41*, 1292–1304. [CrossRef]
34. Cao, L.; Zhu, Y.; Tang, G.; Yuan, F.; Yan, Z. Climatic warming in China according to a homogenized data set from 2419 stations. *Int. J. Climatol.* **2016**, *36*, 4384–4392. [CrossRef]
35. Luo, M.; Lau, N. Amplifying effect of ENSO on heat waves in China. *Clim. Dyn.* **2019**, *52*, 3277–3289. [CrossRef]
36. Yuan, X.; Wang, L.; Wu, P.; Ji, P.; Sheffield, J.; Zhang, M. Anthropogenic shift towards higher risk of flash drought over China. *Nat. Commun.* **2019**, *10*, 4661. [CrossRef] [PubMed]
37. Ying, M.; Zhang, W.; Yu, H.; Lu, X.; Feng, J.; Fan, Y.; Zhu, Y.; Chen, D. An overview of the China Meteorological Administration tropical cyclone database. *J. Atmos. Ocean. Technol.* **2014**, *31*, 287–301. [CrossRef]
38. Lu, X.; Yu, H.; Ying, M.; Zhao, B.; Zhang, S.; Lin, L.; Bai, L.; Wan, R. Western North Pacific tropical cyclone database created by the China Meteorological Administration. *Adv. Atmos. Sci.* **2021**, *38*, 690–699. [CrossRef]
39. Wang, X.; Xiao, F.; Feng, X.; Li, X. Extraction of large-scale urban area information in China using DMSP/OLS nighttime light data. *Remote Sens. Land Resour.* **2013**, *25*, 159–164.
40. Ward, R.D. *The Climates of the United States*; Ginn: Oxford, UK, 1925; pp. 383–395.
41. Schoetter, R.; Cattiaux, J.; Douville, H. Changes of western European heat wave characteristics projected by the CMIP5 ensemble. *Clim. Dyn.* **2015**, *45*, 1601–1616. [CrossRef]
42. Fischer, E.M.; Schär, C. Consistent geographical patterns of changes in high-impact European heatwaves. *Nat. Geosci.* **2010**, *3*, 398–403. [CrossRef]
43. Chen, Y.; Zhai, P. Revisiting summertime hot extremes in China during 1961–2015: Overlooked compound extremes and sig-nificant changes. *Geophys. Res. Lett.* **2017**, *44*, 5096–5103. [CrossRef]
44. Gao, H.; Luo, Y.; Jiang, X.; Zhang, D.L.; Chen, Y.; Wang, Y.; Shen, X. A statistical analysis of extreme hot characteristics and their relationships with urbanization in southern China during 1971–2020. *J. Appl. Meteorol. Clim.* **2021**, *60*, 1301–1317. [CrossRef]
45. Oke, T.R. The energetic basis of the urban heat island. *Quart. J. Roy. Meteor. Soc.* **1982**, *108*, 1–24. [CrossRef]
46. Ding, Y. *Monsoons over China*; Kluwer Academic Publishers: Dordrecht, The Netherlands; Boston, MA, USA; London, UK, 1994; p. 419.
47. Gaffin, D.M. Foehn winds that produced large temperature differences near the southern Appalachian Mountains. *Weather Forecast.* **2007**, *22*, 145–159. [CrossRef]
48. Zhao, D.; Lin, Y.; Li, Y.; Gao, X. An extreme heat event induced by Typhoon Lekima (2019) and its contributing factors. *J. Geophys. Res. Atmos.* **2021**, *126*, e2021JD034760. [CrossRef]
49. Knutson, T.R.; McBride, J.L.; Chan, J.; Emanuel, K.; Holland, G.; Landsea, C.; Held, I.; Kossin, J.P.; Srivastava, A.K.; Sugi, M. Tropical cyclones and climate change. *Nat. Geosci.* **2016**, *3*, 157–163. [CrossRef]
50. Walsh, K.J.E.; McBride, J.L.; Klotzbach, P.J.; Balachandran, S.; Camargo, S.J.; Holland, G.; Knutson, T.R.; Kossin, J.P.; Lee, T.C.; Sobel, A.; et al. Tropical cyclones and climate change. *Wires. Clim. Chang.* **2016**, *7*, 65–89. [CrossRef]
51. Matthews, T.; Wilby, R.L.; Murphy, C. An emerging tropical cyclone-deadly heat compound hazard. *Nat. Clim. Chang.* **2019**, *9*, 602–606. [CrossRef]
52. Provencal, S.; Bergeron, O.; Leduc, R.; Barrette, N. Thermal comfort in Quebec City, Canada: Sensitivity analysis of the UTCI and other popular thermal comfort indices in a mid-latitude continental city. *Int. J. Biometeorol.* **2016**, *60*, 591–603. [CrossRef] [PubMed]

land

MDPI

Article

Long-Term Variations in Warm and Cold Events in Nanjing, China: Roles of Synoptic Weather Patterns and Urbanization

Weishou Tian [1], Lian Zong [1], Yakun Dong [2], Duanyang Liu [3] and Yuanjian Yang [1,*]

1 Collaborative Innovation Centre on Forecast and Evaluation of Meteorological Disasters, School of Atmospheric Physics, Nanjing University of Information Science & Technology, Nanjing 210044, China
2 Nanjing Foreign Language School, Nanjing 210018, China
3 Key Laboratory of Transportation Meteorology of China Meteorological Administration, Nanjing Joint Institute for Atmospheric Sciences, Nanjing 210041, China
* Correspondence: yyj1985@nuist.edu.cn

Abstract: Studying the long-term variations in warm and cold events and their causes under global warming is important for understanding urban climate change, planning, and green development, as well as for disaster prevention and mitigation. In this study, taking the megacity of Nanjing in China as an example, we analyzed the trends and characteristics of the daily average temperature, daily maximum temperature, daily minimum temperature, and warm and cold events from 1960 to 2021, and their association with synoptic weather patterns (SWPs) and urbanization. The results showed that, over the past 62 years, the maximum/average/minimum temperatures in Nanjing have trended upward significantly (at the urban station they increased at rates of 0.17, 0.34, and 0.67 °C/decade), with the minimum temperatures being the most significant. In spring, the warming rate of the average temperature was the greatest, reaching 0.45 °C/decade. All other seasons had their highest warming rate in their minimum temperatures, reaching 0.38 °C/decade, 0.73 °C/decade, and 0.67 °C/decade in the summer, autumn, and winter, respectively. The extreme high temperatures showed a decreasing trend until the mid-1980s, closely related to the decrease in the two SWPs with prevailing southwesterly winds (Types 1 and 2), while a significant increasing trend was apparent thereafter, mainly related to the increase in the SWPs with prevailing southeasterly winds (Types 3 and 4). The number of warm days was strongly positively correlated with extreme high temperatures during the study period, and about 91% of the warm day interannual variation can be explained by extreme high temperature variation. The extreme low temperatures showed a significant decreasing trend. The number of cold nights was strongly and positively correlated with extreme low temperatures, and about 85% of the cold night interannual variation can be explained by extreme low temperature variation. The effect of urbanization was basically positive, contributing the most to the average temperatures and second-most to the minimum temperatures, with warming contributions of 26.5% and 20.9%, respectively, and an insignificant contribution to the maximum temperatures. The effect of urbanization on extreme high temperatures was not significant, but the contribution of warming to extreme low temperatures reached 27.9%. Our results have important implications for future urban climate prediction, as well as for impact assessment and decision making in urban planning.

Keywords: Nanjing; climate change trends; extreme temperature; urbanization; synoptic weather pattern

1. Introduction

The IPCC's Sixth Assessment Report states that the global mean surface temperature has increased by about 1 °C since 1850–1900 and this trend is expected to continue in the near future [1]. Global warming has significant impacts on climate and ecosystems, such as sea level rise, melting of permafrost glaciers, reduced biodiversity, and frequent extreme weather [2,3]; moreover, it has far-reaching impacts on human socioeconomic systems, such as crop yields, energy waste, and human health [4–6]. Many urban areas around the world are experiencing a changing climate and increasing incidence of extreme weather [7–10]. Therefore, studying the long-term changes in hot and cold events and their causes in the context of global warming is important for understanding urban climate change, planning human development, and for disaster prevention and mitigation.

In addition to global climate change, human activities (e.g., urbanization) can also have an impact on urban climate change trends as well as extreme cold and warm events [11–14]. One of the most important features of urbanization is land use and land cover change (replacement of the original natural surface, forests, lakes, grasslands, etc., with roads, bridges, buildings, factories, concrete parking lots, etc.) [15–17]. Urban expansion accompanied by urban population growth alters the original physical properties of the surface (e.g., albedo, roughness, specific heat/moisture coefficient), and coupled with anthropogenic heat emissions, the ability of cities to absorb, store, and emit energy is altered [18,19], leading to the well-known urban heat island effect [20]. It has been noted that the direct effect of urban sprawl on temperature warming can even reach 4 °C [21]; urbanization leads to an additional 0.05 °C/decade of average surface temperature warming in southeastern China [22]. The increase in surface temperature increases the risk of extreme heat. Under the synergistic effects of global warming and local urbanization, rapidly growing urban populations are more vulnerable to the threat of extreme heat [23–25]. Over the past few decades, China, as a populous country with hundreds of millions of urban residents and one of the most vulnerable regions to extreme weather events, has experienced rapid urbanization, which may further exacerbate the long-term trend of extreme temperatures under global warming [26–28]. These trends are particularly evident in highly urbanized and densely populated areas, such as the Yangtze River Delta region [29,30].

As another important driver, different synoptic weather patterns (SWPs) can cause significant temperature changes through modulation of local meteorological factors [1,31–34]. For example, high-pressure systems in summer can suppress the development of the planetary boundary layer and induce calm and cloud-free conditions favorable for radiation enhancement, thus raising temperatures [35–37]. The summer in east-central China is mainly influenced by the East Asian summer wind, which moves from south to north, and the number of high temperature days in southern China increases significantly after the end of the rainy season [38,39]. The western Pacific subtropical high (WPSH) is an important factor in the monsoon system in terms of generating high temperatures in southeastern China [40]. There is significant interannual variability in the extent, intensity, and location of the WPSH, and its positional configuration with the westerly jet and South Asian high (SAH) affects the region where high temperatures occur. For example, the high frequency of high temperature days in the Yangtze River basin in 2013 was closely related to the northwestward position of the WPSH and the strong and northeastward position of the SAH [40].

Nanjing is regarded as one of China's "Four Furnaces", which refers to its especially hot and oppressively humid summer weather within China. The long plum-rain season, a subtropical Asian wet season in the late spring and early summer which causes extreme and consistently high humidity levels, contributes significantly to its hottest temperatures in summer. Nanjing has a resident population of 9.42 million. It is not only one of the fastest growing metropolises in China in recent decades, but also a typical subtropical, inland city [41,42]. However, to date, there have been few previous studies on climate trend changes in Nanjing [43–45], leaving a knowledge gap regarding synoptic-scale changes and their potential drivers. Overall, there are two questions about climate change trends in Nanjing that have not yet been fully answered: (a) What is the contribution of urbanization to long-term climate change trends of cold and warm events? (b) What is the synoptic circulation situation for extreme cold and warm events? Therefore, taking Nanjing as an example, this study focuses on the long-term daily maximum/mean/minimum temperature trends and extreme cold and warm event trends during 1960–2021, and investigates the role of SWPs and urbanization in the changes of these trends based on surface observations and reanalysis data in an integrated manner. The results of this study are expected to provide a scientific reference for climate prediction and urban planning from the perspective of synoptic weather patterns and urbanization in high-density megacities.

2. Data and Methods

2.1. Study Area and Observational Data

Nanjing, located in the middle and lower reaches of the Yangtze River Delta plain, is a metropolis that combines technological modernity with natural beauty. It is located in the East Asian monsoon region, which belongs to the subtropical monsoon climate and is an area sensitive to climate change and prone to meteorological disasters [42]. With global warming in recent years, climate extremes have occurred, but studies on long-term climate change characteristics in Nanjing are incomplete [43,45].

The daily maximum, daily average, and minimum temperatures of the Nanjing National Climate Observatory from 1960 to 2021 were selected for analysis. The observed data were obtained from the China Meteorological Data Service Center (CMDC: http://data.cma.cn/en (accessed on 22 December 2022)). In general, the quality of the raw data is strictly controlled by the CMDC, and additionally the raw data were homogenized using the method proposed by Xu et al. [46]. Furthermore, the hourly geopotential height (GH) and zonal/meridional wind vector were obtained from the Fifth Major Global Reanalysis of the European Centre for Medium-Range Weather Forecasts for further analysis of synoptic weather types (spatial and temporal resolution of 0.25°) [47,48].

2.2. Research Methods

2.2.1. Division of Station Types

Firstly, following previous studies [24,27], a circular buffer zone with a radius of 2 km was constructed around each station. Most of the stations were 2–3 km away from areas of city or county town when they were initially built, and 2 km was chosen because this scale can represent well the local effects of urban expansion, as suggested by many climate studies [13,49,50]. The proportion of the built-up area was calculated based on the corresponding land cover within the buffer zone [24,41]. A station was classified as urban type if it had a built-up area fraction $\geq 25\%$ [27]. The Nanjing station met the criteria for an urban station (although there was a station relocation in 2007), and the Pukou station was identified as a rural reference station; the specific locations are shown in Figure 1.

2.2.2. Definition of Extreme Events and Urbanization Correlation

Based on previous studies [51,52], a weather process in which the daily maximum temperature exceeded 35 °C for three or more consecutive days was defined as a heat wave (HW) day; and a daily minimum temperature below −5 °C was defined as a low temperature (LT) day in this study. The more traditional percentile threshold method was used for warm days (nights) and cold days (nights) [26,53]. Specifically, the maximum (minimum) temperature information of the station during the study period was arranged in ascending order, and the 95th (5th) percentile value was taken as the threshold value. When the maximum temperature of a day was above (below) the threshold, it was considered a warm (cold) day event; when the minimum temperature of a day was above (below) the threshold, it was considered a warm (cold) night event.

The urbanization effect (*Ue*) was defined as the difference in observational temperature trends (°C/decade) between urban and rural stations [24,27], i.e., $Ue = Tdu - Tdr$, in which *Tdu* and *Tdr* are the linear trends (°C/decade) for the observed temperatures of the urban and rural stations, respectively.

Following previous studies [24,27], the contribution of urbanization (*Uc*), the recognized relative impact of urbanization on temperature change, was calculated from *Ue* and *Tdu* (i.e., $Uc = (Ue\ /\ Tdu) \times 100\%$). In this study, simple linear regression was used to calculate the linear trend and the Student's t-test was used to assess the statistical significance of the trend [54]. Correlation coefficients were used to assess the relationship between relative extreme indices and extreme heat waves (low temperatures).

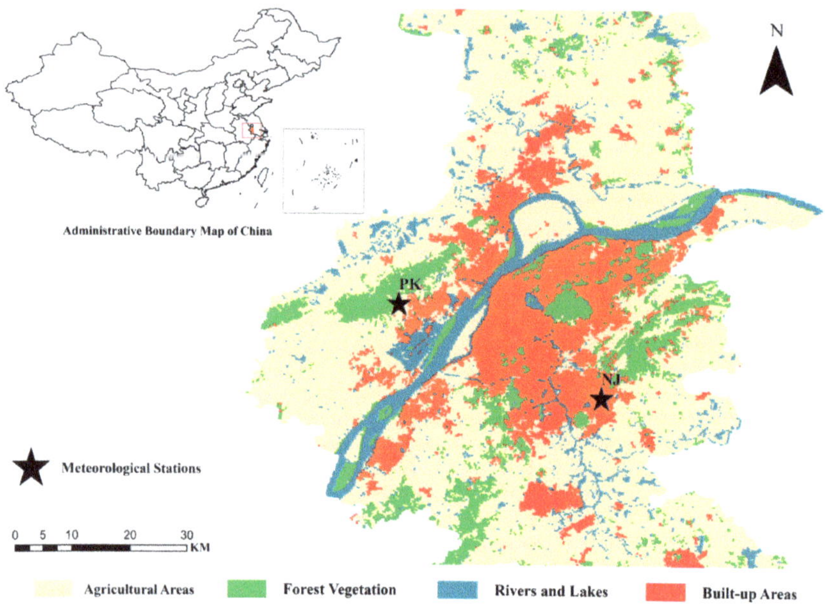

Figure 1. The study area and the locations of the two national meteorological stations in Nanjing, China (map based on the publicly available land use map provided on the website of the Nanjing Municipal Bureau of Land and Resources).

2.2.3. Classification of SWPs

T-mode principal component analysis (T-PCA) is an improved mathematical method for the classification of SWPs [47,55], which can reproduce the preset dominant SWPs without relying on the preset parameters and has the temporal and spatial stability of advanced classification [48,56]. It decomposes the original data matrix into the product of the principal component matrix and the load matrix (two low-dimensional matrices), then rotates the first r ($r \leq n$) principal components with larger variance contributions obliquely, and finally obtains each time the size of the weather pattern and classification according to the loading [51]. In this study, the T-PCA analysis module of the COST733 software (http://cost733.met.no/ (accessed on 5 July 2022)), developed by the European Cooperation for Scientific and Technological Research, was used to classify the number of days of HW or LT occurrence during the study period based on the 850-hPa GH field. To assess the performance of synoptic classification and determine the number of classes, the explained cluster variance method was selected in this study [33,57,58].

3. Results

3.1. Long-Term Trends of Maximum/Average/Minimum Temperatures

Figure 2 shows the evolution of the maximum/average/minimum temperatures at the urban station and the rural reference station in Nanjing from 1960 to 2021. It can be seen that the maximum/average/minimum temperatures show an increasing trend during the study period, and most of these indicators are statistically significant. The maximum temperature is the most direct indicator of the heat condition. The annual maximum temperatures at the urban and rural stations show an increasing trend over the last 62 years. The rate of change was greater at the rural station than at the urban station, at 0.31 °C/decade and 0.17 °C/decade, respectively. Although the increase at the urban station does not pass the 0.05 significance test, the greatest increase in warming was in the spring in both cases (Figure S1). The warming trend of the average temperatures at the urban station was significantly greater than that at the rural station (0.25 °C/decade), at approximately 0.34 °C/decade. The warming rate at both station types was greatest in the spring, followed by winter, and the average temperature warming in summer was slower, or even does not increase, compared to other seasons. It is noteworthy that the warming of minimum temperatures at both the urban and rural stations was highly significant, with rates of change of 0.67 °C/decade and 0.53 °C/decade, respectively. The warming of minimum temperatures was greatest in autumn and winter, and the rate of change at the urban station was significantly greater than that at the rural station. In summary, the increase in minimum temperatures at the urban and rural sites was much greater than for the maximum and average temperatures, indicating that the warming in Nanjing was mainly reflected in the increase in minimum temperatures, in addition to the gradual warming of Nanjing in autumn and winter.

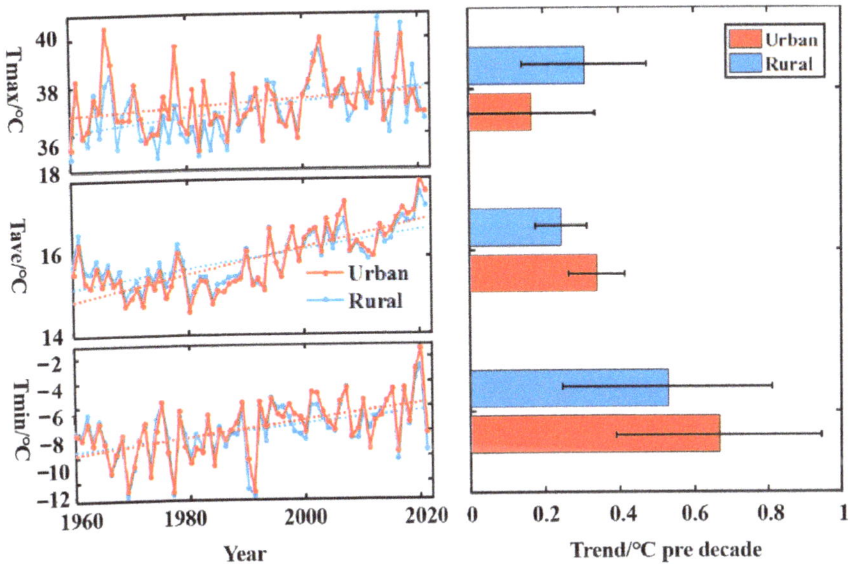

Figure 2. Time series for maximum/mean/minimum temperatures in urban areas and rural areas during 1960–2021. The red and blue dashed lines in the left-hand panel and the columns in the right-hand panel indicate their corresponding linear trends. Error bars in the right-hand panels indicate the 95% confidence interval.

3.2. *Urbanization Effects*

From the above analysis, it can be seen that the trends of maximum/average/minimum temperatures at the urban and rural stations were basically the same, with an increasing trend. In addition, the urban station had a more significant upward trend than the rural station in terms of mean and minimum temperatures. In order to better understand the influence of urbanization on the temperature trends in Nanjing, the impact of urbanization on maximum/mean/minimum temperatures and its contribution were analyzed (Table 1). On the annual scale, the effect of urbanization on maximum temperatures was not significant. On the seasonal scale, urbanization passed the 0.05 significance test only in spring, contributing 25% to the warming of the maximum temperatures. On the annual scale, the impact of urbanization on the average temperatures was more significant, with a contribution of 26.5%. On the seasonal scale, except for summer, when the contribution of urbanization to the average temperatures did not pass the significance test, for all other seasons it was positive, with autumn being the highest, followed by winter and spring, at 39.2%, 19%, and 17.8%, respectively. On the annual scale, the effect of urbanization on the minimum temperatures was significant, with a contribution rate of 20.9%. On the seasonal scale, the highest value for the urbanization contribution to minimum temperatures was in autumn, followed by summer and winter, and the lowest value was in spring, with contributions of 30.1%, 28.9%, 20.9%, and 18.8%, respectively, all of which passed the 0.05 significance test. Overall, the effect of urbanization was basically positive, contributing the most to the average temperatures, and second-most to the minimum temperatures, with an insignificant contribution to the maximum temperatures. Urbanization contributed the most to the maximum temperatures in spring, the most to average temperatures in autumn, and the most to minimum temperatures in summer and winter.

Table 1. Interannual and intraseasonal trends in urban and rural maximum/mean/minimum temperatures, along with the contribution of urbanization (statistical significance level at the 0.05 is denoted by *, and a tilde (~) indicates that the effect of urbanization failed the 0.05 significance test and no calculation of the urbanization contribution was performed).

Time	Urban Max/Mean/Min Temperature Trends (°C/Decade)	Rural Max/Mean/Min Temperature Trends (°C/Decade)	Urbanization Contribution Rate (%)
Interannual variation	0.17/0.34 */0.67 *	0.31 */0.25 */0.53 *	~/26.5/20.9
Spring	0.32 */0.45 */0.32 *	0.24 */0.37 */0.26 *	25/17.8/18.8
Summer	0.17/0.11/0.38 *	0.3 */0.02/0.27 *	~/~/28.9
Autumn	−0.03/0.33 */0.73 *	0.14/0.2 */0.51 *	~/39.4/30.1
Winter	0.12/0.42 */0.67 *	0.11/0.34 */0.53 *	~/19/20.9

3.3. *SWPs Corresponding to Extreme Warm and Cold Events*

3.3.1. Changes in HW Events and SWPs

Figure 3 shows the 62-year time series for HW days at the urban and rural stations in Nanjing. It can be seen that, before the mid-1980s, the number of HW days shows a significant decreasing trend and the rate of change is slightly higher in urban than rural areas, at −2.4 d/decade and −2.2 d/decade, respectively. However, after the mid-1980s, a significant increasing trend can be seen and the rate of change is slightly higher at the rural than at the urban station, at 0.19 d/decade and 0.16 d/decade, respectively. This is similar to previous reports of extreme maximum temperatures showing a significant warming trend after the 1990s [53,59].

To further elucidate the synoptic weather background during the occurrence of HW days in Nanjing, four favorable weather patterns were identified (Figure 4): (1) Type 1, in which the center of the WPSH is located to the southeast of Nanjing, and Nanjing is under its control, with strong cyclone appearing to the north of Nanjing and prevailing southwesterly winds (Figure 4a); (2) Type 2, which is controlled by the WPSH, with the center of anticyclone located to the east of Nanjing, resulting in prevailing southwesterly winds over Nanjing (Figure 4b); (3) Type 3, which comprises two types of weaker anticyclones appearing in the southeast and northwest of Nanjing, with Nanjing mainly under the control of the southeast anticyclone and a strong cyclone system in the northeast of Nanjing, resulting in very weak southeasterly winds prevailing in Nanjing with almost zero wind speed (Figure 4c); and (4) Type 4, which features a strong WPSH located northeast of Nanjing, controlling the entire area of Nanjing, and weak cyclone located south of Nanjing, leading to prevailing southeasterly winds (Figure 4d). The variability of the prevailing atmospheric circulation system may be directly responsible for significant increases in temperature [43]. Overall, the occurrence of HW days in Nanjing is mainly related to the WPSH (with southerly winds prevailing), and the synoptic weather types favoring the occurrence of HW events follow the order Type 1 (35.15%) > Type 2 (21.87%) > Type 4 (4.84%) > Type 3 (4.21 %).

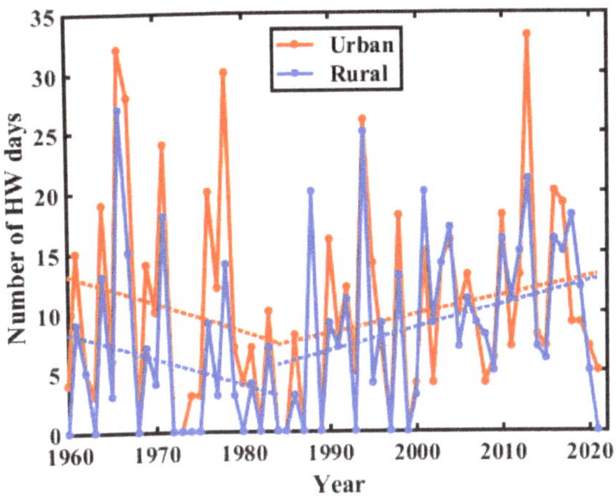

Figure 3. Changes and trends of urban heat wave (HW) days (orange) and rural HW days (blue) from 1960 to 2021.

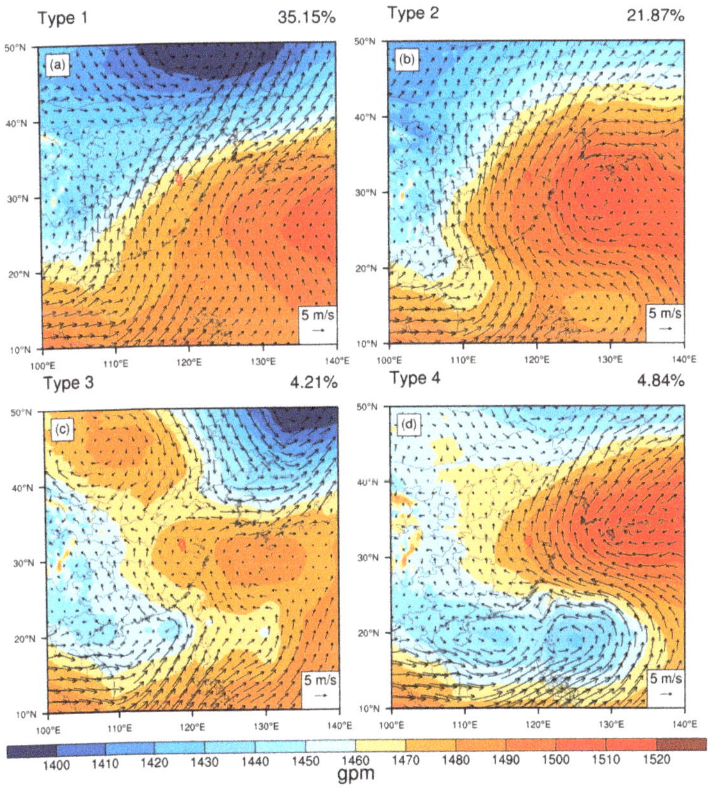

Figure 4. The 850-hHPa geopotential height (contours) and zonal/meridional wind (vector) patterns associated with the occurrence of HW events over 62 years at 1400 LST (local standard time) based on objective classification (the red outline indicates Nanjing and the number in the upper-right corner of each panel indicates the frequency of occurrence for each pattern).

In order to explore the regulation of HW days by different SWPs, we investigated the variation in the frequency of HW days under the four different synoptic weather types mentioned above (Figure 5). The patterns of variation in HW days under the four weather types were basically similar. However, the decreasing trend of Type 1 and Type 2 was closer to the trend of change in HW days before the mid-1980s (Figure 3); while the increasing trend of Type 3 and Type 4 was more significant after the mid-1980s, which is closely consistent with the trend of change in HW days in that period.

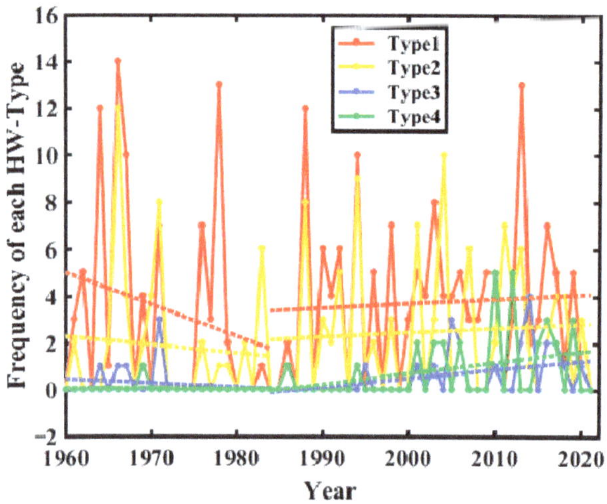

Figure 5. Changes in heat wave (HW) day frequency corresponding to each of the four synoptic weather types.

In conclusion, the frequency of extreme HW days declined up until the mid-1980s, which was closely related to the decline in the Type 1 and Type 2 SWPs, and urbanization had no significant effect. However, there was a significant upward trend in HW days after the mid-1980s and this was related to an increase in the Type 3 and Type 4 SWPs, and urbanization again had no significant effect.

3.3.2. Changes in LT Events and SWPs

Figure 6 shows the 62-year time series for LT days at the urban and rural stations in Nanjing. The trend in extreme LT days was a continuously decreasing one, and the decreasing trend at the urban station was faster than that at the rural station, at 1.8 d/decade and 1.3 d/decade, respectively. This confirms the above conclusion that the warming of the Nanjing area was precisely reflected by the decrease in LT days.

To further elucidate the synoptic weather background for the occurrence of LT days in Nanjing, four favorable weather patterns were identified (Figure 7): (1) Type 1, in which the Mongolian anticyclone center is located in the northwestern part of Nanjing and the northeastern part of Nanjing is controlled by a strong northeastern cold cyclone system with a northwesterly jet over Nanjing under the convergence of anticyclone and cyclone systems, and prevailing northwesterly winds (Figure 7a); (2) Type 2, in which a anticyclone center is located in the southwest of Nanjing with mainly northwesterly winds (Figure 7b); (3) Type 3, with a weaker anticyclone center located in the south of Nanjing leading to prevailing westerly winds over Nanjing (Figure 7c); (4) Type 4, with an anticyclone center located in the west of Nanjing and a weak cyclone system in the northeast near Nanjing, also with a northwesterly jet leading to strong northwesterly winds over Nanjing (Figure 7d). Overall, the occurrence of LT days in Nanjing is mainly due to the control of cold anticyclone (prevailing northerly winds), and the weather types favoring the occurrence of LT days follow the order Type 2 (21.25%) > Type 1 (19.84%) > Type 3 (11.25%) > Type 4 (10.00%).

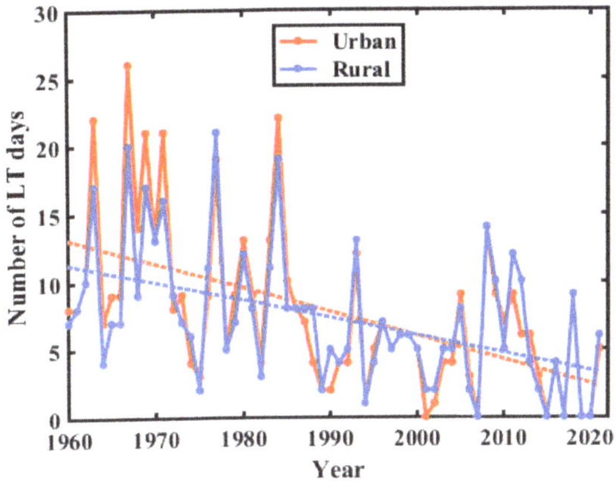

Figure 6. Changes and trends in urban low temperature (LT) days (orange) and rural LT days (blue) from 1960 to 2021.

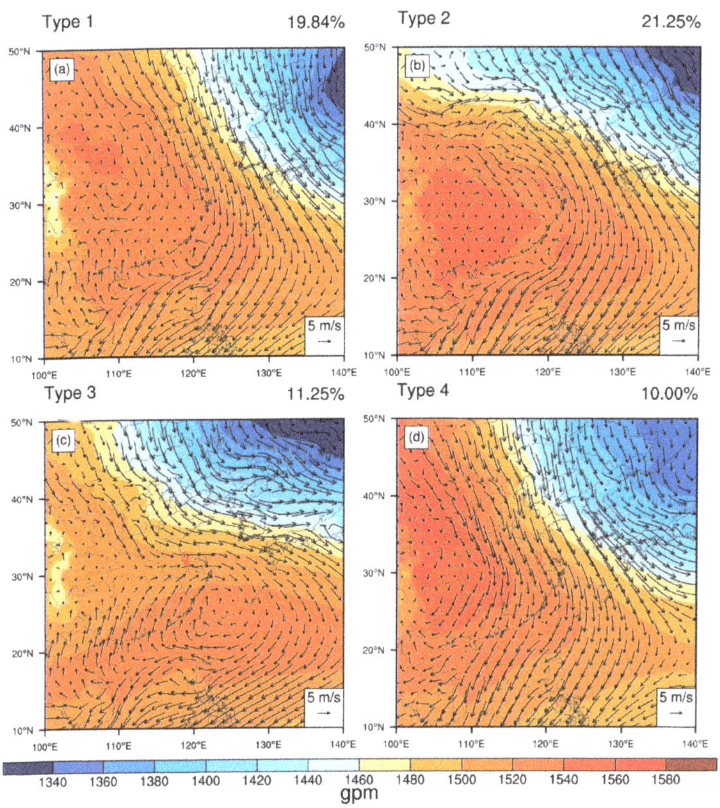

Figure 7. The 850-hHPa geopotential height (contours) and zonal/meridional wind (vector) patterns associated with the occurrence of LT events over 62 years at 0200 LST (local standard time) based on objective classification (the red outline indicates Nanjing and the number in the upper-right corner of each panel indicates the frequency of occurrence for each pattern).

To explore the regulation of LT days by different SWPs, we investigated the variation in the frequency of LT days under the four different synoptic weather types mentioned above (Figure 8). The patterns of variation in LT days under the four SWPs were basically similar. Notably, the decreasing trend in Type 2 was closer to the trend in LT days.

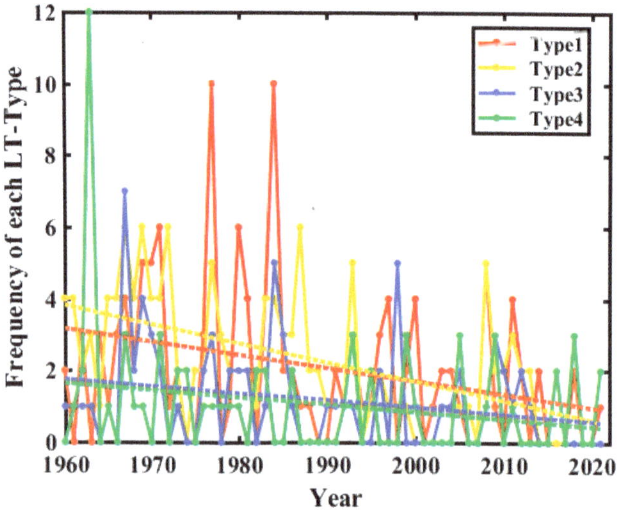

Figure 8. Changes in the frequency of low temperature (LT) days corresponding to each of the four synoptic weather types.

In conclusion, the frequency of extreme LT days decreased during the study period, a trend that was closest to that of the Type 2 SWP. Also, it is noteworthy that urbanization had a significant impact on it, with a contribution of 27.85%.

3.4. Changes in Relative Extreme Temperature Indices

The relative extremes of the maximum temperatures include the number of warm days and the number of cold days. Figure 9a shows the trend of warm days in urban and rural areas with an increase of 0.45 d/decade and 1.45 d/decade, respectively (Table S1). The rate of increase in warm days in urban areas was significantly lower than that in rural areas, but did not pass the 0.05 significance test, while the increasing trend in rural areas passed. This increasing trend occurred mainly after the mid-1980s, which is closely consistent with the trend in extreme HW days (Figure 3). Figure 9b shows the trend in the frequency of cold days in urban and rural areas. The decreasing rate of cold days in urban areas was 1.32 d/decade. The reduction rate of 1.15d/decade in rural areas was lower than the reduction rate in urban areas and did not pass the 0.05 significance test (Table S1). The relative extremes of the minimum temperatures include changes in the number of warm nights and cold nights. Figure 9c shows the trend in the number of warm nights in urban and rural areas, from which it can be seen that the number of warm nights was increasing, with increases of 1.74 d/decade and 1.82 d/decade, respectively (Table S1), both passing the 0.05 significance test. Similarly, there was no significant trend in warm nights before the mid-1980s, while a trend of increasing warm nights was clear after that. The significant increase in the frequency of warm nights in Nanjing after the mid-1980s may be related to its rapid economic development and industrialization after China's "reform and opening up" period. Figure 9d shows that the number of cold nights in urban and rural areas had a clear decreasing trend, with a decrease rate of 3.18 d/decade and 2.48 d/decade, respectively, both passing the 0.05 significance test (Table S1). This is closely consistent with the variation in extreme LT days (Figure 6).

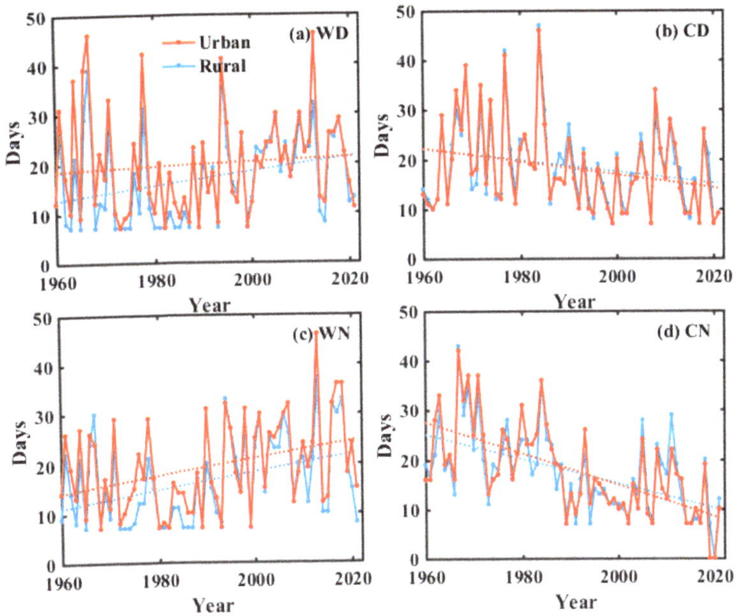

Figure 9. Changes in relative extreme temperature indices in urban/rural areas in Nanjing during 1960–2021: (**a**) warm days; (**b**) cold days; (**c**) warm nights; (**d**) cold nights.

4. Discussion

Previous studies have shown that extreme events have a tendency to spread more intensely, frequently, and destructively in urban areas [7,8,10]. In recent years, studies on extreme events and their associated drivers in China have received increasing attention [11,12,22]. Especially in the context of global warming and still accelerating human activities, the impact of urbanization on regional trends of mean, maximum, and minimum temperatures has been widely analyzed [14,26]. Although the trends vary in different regions, the vast majority of studies show that there is a clear asymmetry in the changes of maximum and minimum temperatures, with the rising trend of minimum temperatures being more pronounced than that of maximum temperatures [24,53,54]. This is generally consistent with the results of our study. However, the difference is that in our study urbanization contributed the most to the average temperature, and the second most to the minimum temperature, with warming contributions of 26.5% and 20.9%, respectively, while the contribution to the maximum temperature did not pass the significance test. Similarly, the contribution of urbanization to extreme high temperature did not pass the significance test, but the contribution to extreme low temperature reached 27.9%.

From a seasonal perspective, urbanization made the greatest contribution to the maximum temperature in spring, to the average temperature in autumn, and to the minimum temperature in summer and winter. This differs from previous studies [24,54], which may be due to climate differences caused by regional differences in urbanization or differences in station selection and classification. In addition, compared to previous studies [40,46,53], we further explored the contribution of large-scale circulation in urban climate extremes. The occurrence of extreme heat waves in Nanjing is mainly related to WPSH (prevailing southerly winds), and the occurrence of extreme low temperatures is mainly due to the control of cold anticyclones (prevailing northerly winds), and different types of weather types contribute differently (specific results are presented above and will not be repeated here).

Relative extreme temperature indices have changed significantly in many regions of the world over the past half-century [24,26,60,61]. These changes are closely related to the changes in extreme HW and LT events, as both relative extreme temperature indices depend on daily maximum and minimum temperature records. In the case of the urban station, the two indices of warm days and HW days were strongly positively correlated (correlation coefficient $r = 0.91$) (Table 2). After removing the linear trend, the correlation was still strong ($r = 0.90$, >99% confidence level), indicating that the annual variability of warm days was closely related to the variability of HW days. Cold days and LT days showed a strong positive correlation ($r = 0.89$). After removing the linear trend, the correlation was still strong ($r = 0.84$, >99% confidence level), indicating that the annual variability of cold days was closely related to that of LT days. Similarly, warm nights and cold nights were closely correlated with HW and LT days, respectively.

Table 2. Correlation and significance testing between relative extreme indexes and HW (LT).

	Urban	Rural
warm day–HW	$r = 0.91, p < 0.01$	$r = 0.86, p < 0.01$
cold night–LT	$r = 0.89, p < 0.01$	$r = 0.88, p < 0.01$
warm night–HW	$r = 0.75, p < 0.01$	$r = 0.79, p < 0.01$
cold day–LT	$r = 0.61, p < 0.01$	$r = 0.60, p < 0.01$

Global warming means not only an increase in global temperature but also a change in the climate system, the most direct impact of which leads to extreme events such as HWs, LTs, floods, and droughts. The health risks associated with extreme weather threaten the fate and health of all mankind [2,3,13,51]. For instance, in summer of 2022, the average temperature in mid-eastern China was the highest since 1961 and the average precipitation was the second lowest controlled by an anomalous high pressure system, with 15.1% of national weather stations recording daily maximum temperatures that were equal to or exceeded recorded historical extremes, together causing problems for electricity supply, drought conditions, cases of heat stroke, and even deaths [62,63]. The impact of ambient temperature on residential mortality was systematically assessed at the national level in a study covering 272 major cities in China and collecting data on 1,826,000 non-accidental deaths [64]. From the perspective of interannual variation in synoptic patterns, therefore, understanding long-term changes in extreme cold and warm events is important for the detection and attribution of climate change and human health adaptation in cities, especially for a densely populated city.

5. Conclusions

This study investigated the long-term maximum/mean/minimum temperature trends and extreme cold and warm event trends from 1960–2021 in Nanjing, China, as well as the role of SWPs and urbanization in the changes in these trends, based on surface observations and reanalysis data. The following conclusions were obtained.

Over the past 62 years, the maximum/mean/minimum temperatures in Nanjing showed a significant upward trend (0.17, 0.34, and 0.67 °C/decade for the urban station; 0.31, 0.25, and 0.53 °C/decade for the rural station), with the minimum temperature warming rates being the most significant. In spring, the greatest warming rate of 0.45 °C/decade was observed for the mean temperature. In summer, autumn, and winter, the warming rate of the minimum temperatures was the largest, reaching 0.38 °C/decade, 0.73 °C/decade, and 0.67 °C/decade, respectively. This indicates that the warming in Nanjing was more obvious in the minimum temperatures, and the Nanjing area was gradually warming in autumn and winter.

The extreme high temperatures showed a decreasing trend until the mid-1980s, closely related to the decrease in the two SWPs with prevailing southwesterly winds (Types 1 and 2). Then, a significant increasing trend was observed, mainly related to the increase in SWPs with prevailing southeasterly winds (Types 3 and 4). The number of warm days was strongly positively correlated with extreme high temperatures during the study period, and about 91% of the warm day interannual variation can be explained by extreme high temperature variation. The frequency of extreme LT days showed a significant decreasing trend. The number of cold nights was strongly and positively correlated with extreme LTs, and about 85% of the cold night interannual variation can be explained by extreme LT variation. The relative extreme indexes of warm days and warm nights generally showed an upward trend, while cold days and cold nights showed a downward trend. Note that the trends in warm days and warm nights remained closely consistent with that of HW days, and, likewise, the trends in cold days and cold nights were also highly consistent with that of LT days.

In general, our findings provide a useful scientific reference for climate prediction, urban planning, and human development from the perspective of synoptic weather patterns and urbanization in large, high-density cities.

Supplementary Materials: The following supporting information can be downloaded at: https://www.mdpi.com/article/10.3390/land12010162/s1. Figure S1. Time series of maximum/mean/minimum temperature in urban areas and rural areas during 1960–2021; Table S1. Interannual and intraseasonal trends in urban and rural warm days; cold days; warm nights; cold nights, along with the contribution of urbanization.

Author Contributions: Conceptualization, Y.Y.; methodology, W.T. and Y.Y.; data curation, D.L. and W.T.; software, W.T.; validation, W.T. and Y.Y.; formal analysis, W.T., L.Z., Y.D., D.L. and Y.Y.; writing—original draft preparation, W.T.; writing—review and editing, W.T., L.Z., Y.D., D.L. and Y.Y.; visualization, W.T.; supervision, Y.Y.; funding acquisition, Y.Y. All authors have read and agreed to the published version of the manuscript.

Funding: This study is supported by the National Natural Science Foundation of China (42222503 and 42175098).

Institutional Review Board Statement: Not applicable.

Informed Consent Statement: Not applicable.

Data Availability Statement: Please direct any inquiries regarding the data to the corresponding author (yyj1985@nuist.edu.cn).

Acknowledgments: We thank all the scientists, engineers, and students who participated in the field experiments, maintained the instruments, and processed the measurements.

Conflicts of Interest: The authors declare no conflict of interest.

References

1. IPCC. *Climate Change 2021: The Physical Science Basis. Contribution of Working Group I to the Sixth Assessment Report of the Intergovernmental Panel on Climate Change*; Masson-Delmotte, V., Zhai, P., Pirani, A., Connors, S., Péan, C., Berger, S., Caud, N., Chen, Y., Goldfarb, L., Gomis, M., et al., Eds.; Cambridge University Press: Cambridge, UK; New York, NY, USA, 2021; *in press*.
2. AghaKouchak, A.; Cheng, L.; Mazdiyasni, O.; Farahmand, A. Global warming and changes in risk of concurrent climate extremes: Insights from the 2014 California drought. *Geophys. Res. Lett.* **2014**, *41*, 8847–8852. [CrossRef]
3. Gasparrini, A.; Guo, Y.; Hashizume, M.; Lavigne, E.; Zanobetti, A.; Schwartz, J.; Tobias, A.; Tong, S.; Rocklöv, J.; Forsberg, B.; et al. Mortality risk attributable to high and low ambient temperature: A multicountry observational study. *Lancet* **2015**, *386*, 369–375. [CrossRef] [PubMed]
4. Tao, F.; Yokozawa, M.; Liu, J.; Zhang, Z. Climate–crop yield relationships at provincial scales in China and the impacts of recent climate trends. *Clim. Res.* **2008**, *38*, 83–94. [CrossRef]
5. Kovats, S.; Akhtar, R. Climate, climate change and human health in Asian cities. *Environ. Urban.* **2008**, *20*, 165–175. [CrossRef]
6. Hondula, D.M.; Balling, R.C.; Vanos, J.K.; Georgescu, M. Rising Temperatures, Human Health, and the Role of Adaptation. *Curr. Clim. Change Rep.* **2015**, *1*, 144–154. [CrossRef]
7. Trenberth, K.E.; Fasullo, J.T. Climate extremes and climate change: The Russian heat wave and other climate extremes of 2010. *J. Geophys. Res. Atmos.* **2012**, *117*, 17103. [CrossRef]
8. Mukherjee, S.; Mishra, A.K. Increase in Compound Drought and Heatwaves in a Warming World. *Geophys. Res. Lett.* **2021**, *48*, e2020GL090617. [CrossRef]
9. Vose, R.S.; Easterling, D.R.; Gleason, B. Maximum and minimum temperature trends for the globe: An update through 2004. *Geophys. Res. Lett.* **2005**, *32*, 1–5. [CrossRef]
10. Meehl, G.A.; Tebaldi, C. More Intense, More Frequent, and Longer Lasting Heat Waves in the 21st Century. *Science* **2004**, *305*, 994–997. [CrossRef]

11. Ren, G.; Zhou, Y.; Chu, Z.; Zhou, J.; Zhang, A.; Guo, J.; Liu, X. Urbanization Effects on Observed Surface Air Temperature Trends in North China. *J. Clim.* **2008**, *21*, 1333–1348. [CrossRef]
12. Luo, M.; Lau, N.-C. Heat Waves in Southern China: Synoptic Behavior, Long-Term Change, and Urbanization Effects. *J. Clim.* **2017**, *30*, 703–720. [CrossRef]
13. Luo, M.; Lau, N. Increasing Human-Perceived Heat Stress Risks Exacerbated by Urbanization in China: A Comparative Study Based on Multiple Metrics. *Earth's Future* **2021**, *9*, e2020EF001848. [CrossRef]
14. Zhang, P.; Ren, G.; Qin, Y.; Zhai, Y.; Zhai, T.; Tysa, S.K.; Xue, X.; Yang, G.; Sun, X. Urbanization Effects on Estimates of Global Trends in Mean and Extreme Air Temperature. *J. Clim.* **2021**, *34*, 1923–1945. [CrossRef]
15. Kalnay, E.; Cai, M. Impact of urbanization and land-use change on climate. *Nature* **2003**, *423*, 528–531, Erratum in *Nature* **2003**, *425*, 102. [CrossRef] [PubMed]
16. Hua, W.; Chen, H.; Li, X. Effects of future land use change on the regional climate in China. *Sci. China Earth Sci.* **2015**, *58*, 1840–1848. [CrossRef]
17. Stewart, I.D.; Oke, T.R.; Krayenhoff, E.S. Evaluation of the 'local climate zone' scheme using temperature observations and model simulations. *Int. J. Climatol.* **2014**, *34*, 1062–1080. [CrossRef]
18. Li, D.; Sun, T.; Liu, M.; Yang, L.; Wang, L.; Gao, Z. Contrasting responses of urban and rural surface energy budgets to heat waves explain synergies between urban heat islands and heat waves. *Environ. Res. Lett.* **2015**, *10*, 054009. [CrossRef]
19. Oke, T.R.; Mills, G.; Christen, A.; Voogt, J.A. *Urban Climates*; Cambridge University Press: Cambridge, UK, 2017.
20. Oke, T.R. The energetic basis of the urban heat island. *Q. J. R. Meteorol. Soc.* **1982**, *108*, 1–24. [CrossRef]
21. Georgescu, M.; Moustaoui, M.; Mahalov, A.; Dudhia, J. Summer-time climate impacts of projected megapolitan expansion in Arizona. *Nat. Clim. Change* **2013**, *3*, 37–41. [CrossRef]
22. Zhou, L.; Dickinson, R.E.; Tian, Y.; Fang, J.; Li, Q.; Kaufmann, R.K.; Tucker, C.J.; Myneni, R.B. Evidence for a significant urbanization effect on climate in China. *Proc. Natl. Acad. Sci. USA* **2004**, *101*, 9540–9544. [CrossRef]
23. Yang, X.; Leung, L.R.; Zhao, N.; Zhao, C.; Qian, Y.; Hu, K.; Liu, X.; Chen, B. Contribution of urbanization to the increase of extreme heat events in an urban agglomeration in east China. *Geophys. Res. Lett.* **2017**, *44*, 6940–6950. [CrossRef]
24. Ren, G.; Zhou, Y. Urbanization Effect on Trends of Extreme Temperature Indices of National Stations over Mainland China, 1961–2008. *J. Clim.* **2014**, *27*, 2340–2360. [CrossRef]
25. Zong, L.; Liu, S.; Yang, Y.; Ren, G.; Yu, M.; Zhang, Y.; Li, Y. Synergistic Influence of Local Climate Zones and Wind Speeds on the Urban Heat Island and Heat Waves in the Megacity of Beijing, China. *Front. Earth Sci.* **2021**, *9*, 673786. [CrossRef]
26. Zhai, P.; Pan, X. Trends in temperature extremes during 1951–1999 in China. *Geophys. Res. Lett.* **2003**, *30*(17), 1913. [CrossRef]
27. Yang, Y.-J.; Wu, B.-W.; Shi, C.-E.; Zhang, J.-H.; Li, Y.-B.; Tang, W.-A.; Wen, H.-Y.; Zhang, H.-Q.; Shi, T. Impacts of Urbanization and Station-relocation on Surface Air Temperature Series in Anhui Province, China. *Pure Appl. Geophys.* **2012**, *170*, 1969–1983. [CrossRef]
28. Zhang, Y.; Ning, G.; Chen, S.; Yang, Y. Impact of Rapid Urban Sprawl on the Local Meteorological Observational Environment Based on Remote Sensing Images and GIS Technology. *Remote Sens.* **2021**, *13*, 2624. [CrossRef]
29. Ren, G.-Y. Urbanization as a major driver of urban climate change. *Adv. Clim. Change Res.* **2015**, *6*, 1–6. [CrossRef]
30. Luo, M.; Lau, N.C. Urban Expansion and Drying Climate in an Urban Agglomeration of East China. *Geophys. Res. Lett.* **2019**, *46*, 6868–6877. [CrossRef]
31. Lu, R. Interannual Variability of the Summertime North Pacific Subtropical High and its Relation to Atmospheric Convection over the Warm Pool. *J. Meteorol. Soc. Jpn. Ser. II* **2001**, *79*, 771–783. [CrossRef]
32. You, Q.; Kang, S.; Aguilar, E.; Pepin, N.; Flügel, W.-A.; Yan, Y.; Xu, Y.; Zhang, Y.; Huang, J. Changes in daily climate extremes in China and their connection to the large scale atmospheric circulation during 1961–2003. *Clim. Dyn.* **2010**, *36*, 2399–2417. [CrossRef]
33. Hoffmann, P.; Schlünzen, K.H. Weather Pattern Classification to Represent the Urban Heat Island in Present and Future Climate. *J. Appl. Meteorol. Clim.* **2013**, *52*, 2699–2714. [CrossRef]
34. Stewart, I.D.; Oke, T.R. Local Climate Zones for Urban Temperature Studies. *Bull. Am. Meteorol. Soc.* **2012**, *93*, 1879–1900. [CrossRef]
35. Miao, Y.; Guo, J.; Liu, S.; Liu, H.; Li, Z.; Zhang, W.; Zhai, P. Classification of summertime synoptic patterns in Beijing and their associations with boundary layer structure affecting aerosol pollution. *Atmos. Chem. Phys.* **2017**, *17*, 3097–3110. [CrossRef]
36. Yang, Y.; Zheng, X.; Gao, Z.; Wang, H.; Wang, T.; Li, Y.; Lau, G.N.C.; Yim, S.H.L. Long-Term Trends of Persistent Synoptic Circulation Events in Planetary Boundary Layer and Their Relationships with Haze Pollution in Winter Half Year Over Eastern China. *J. Geophys. Res. Atmos.* **2018**, *123*, 10,991–11,007. [CrossRef]
37. Wang, P.; Tang, J.; Sun, X.; Wang, S.; Wu, J.; Dong, X.; Fang, J. Heat Waves in China: Definitions, Leading Patterns, and Connections to Large-Scale Atmospheric Circulation and SSTs. *J. Geophys. Res. Atmos.* **2017**, *122*, 10,679–10,699. [CrossRef]
38. Zhou, T.; Wu, B.; Dong, L. Advances in research of ENSO changes and the associated impacts on Asian-Pacific climate. *Asia-Pac. J. Atmos. Sci.* **2014**, *50*, 405–422. [CrossRef]
39. Lu, R.-Y.; Chen, R.-D. A review of recent studies on extreme heat in China. *Atmos. Ocean. Sci. Lett.* **2015**, *9*, 114–121. [CrossRef]
40. Wang, W.; Zhou, W.; Li, X.; Wang, X.; Wang, D. Synoptic-scale characteristics and atmospheric controls of summer heat waves in China. *Clim. Dyn.* **2015**, *46*, 2923–2941. [CrossRef]
41. Chen, S.; Yang, Y.; Deng, F.; Zhang, Y.; Liu, D.; Liu, C.; Gao, Z. A high-resolution monitoring approach of canopy urban heat island using a random forest model and multi-platform observations. *Atmos. Meas. Tech.* **2022**, *15*, 735–756. [CrossRef]
42. Yang, X.; Yao, L.; Jin, T.; Peng, L.L.; Jiang, Z.; Hu, Z.; Ye, Y. Assessing the thermal behavior of different local climate zones in the Nanjing metropolis, China. *Build. Environ.* **2018**, *137*, 171–184. [CrossRef]

43. Tang, G.L.; Ding, Y.H. The Changes in Temperature and Its Possible Causes in Nanjing in Recent 44 Years. *Atmos. Sci.* **2006**, *30*, 13. (In Chinese)
44. Yang, Y.; Jiang, N. Air temperature and heat island effect change Character in Nanjing city for last 50a. *Meteorol. Sci.* **2009**, *29*, 88–91. (In Chinese)
45. Deng, S.; Lu, X.; Lu, B.; Zhang, H.; Zhao, C.; Xu, K.; Mao, D.; Cui, D.; Wu, X. Variation Analysis of Annual Mean Temperature and Precipitation Near 53 Years in Nanjing City. *Hydropower Energy Sci.* **2014**, *32*, 14–17. (In Chinese)
46. Xu, W.; Li, Q.; Wang, X.L.; Yang, S.; Cao, L.; Feng, Y. Homogenization of Chinese daily surface air temperatures and analysis of trends in the extreme temperature indices. *J. Geophys. Res. Atmos.* **2013**, *118*, 9708–9720. [CrossRef]
47. Zong, L.; Yang, Y.; Gao, M.; Wang, H.; Wang, P.; Zhang, H.; Wang, L.; Ning, G.; Liu, C.; Li, Y.; et al. Large-scale synoptic drivers of co-occurring summertime ozone and PM2.5 pollution in eastern China. *Atmos. Chem. Phys.* **2021**, *21*, 9105–9124. [CrossRef]
48. Yang, Y.; Guo, M.; Ren, G.; Liu, S.; Zong, L.; Zhang, Y.; Zheng, Z.; Miao, Y.; Zhang, Y. Modulation of Wintertime Canopy Urban Heat Island (CUHI) Intensity in Beijing by Synoptic Weather Pattern in Planetary Boundary Layer. *J. Geophys. Res. Atmos.* **2022**, *127*, e2021JD035988. [CrossRef]
49. Yang, Y.; Zheng, Z.; Yim, S.Y.; Roth, M.; Ren, G.; Gao, Z.; Wang, T.; Li, Q.; Shi, C.; Ning, G.; et al. PM 2.5 Pollution Modulates Wintertime Urban Heat Island Intensity in the Beijing-Tianjin-Hebei Megalopolis, China. *Geophys. Res. Lett.* **2020**, *47*, e2019GL084288. [CrossRef]
50. Shi, T.; Yang, Y.; Sun, D.; Huang, Y.; Shi, C. Influence of Changes in Meteorological Observational Environment on Urbanization Bias in Surface Air Temperature: A Review. *Front. Clim.* **2022**, *3*, 189. [CrossRef]
51. Zong, L.; Yang, Y.; Xia, H.; Gao, M.; Sun, Z.; Zheng, Z.; Li, X.; Ning, G.; Li, Y.; Lolli, S. Joint occurrence of heatwaves and ozone pollution and increased health risks in Beijing, China: Role of synoptic weather pattern and urbanization. *Atmos. Chem. Phys.* **2022**, *22*, 6523–6538. [CrossRef]
52. Pan, W.; Miao, Q.; Xu, Y. Characteristics of temperature change in Nanjing from 1951 to 2006. *J. Nanjing Meteorol. Inst.* **2008**, *31*, 694–701. (In Chinese)
53. Hua, L.; Ma, Z.; Zeng, Z. The Comparative Analysis of the Changes of Extreme Temperature and Extreme Diurnal Temperature Range of Large Cities and Small Towns in Eastern China. *Atmos. Sci.* **2006**, *30*, 80.
54. Rong, W.; Yi, S.; Yuanjian, Y.; Wusan, X.; Yin, T.; Hao, Z.; Tao, S. Effects of Urbanization on Extreme Temperature Events in Anhui Province. *Adv. Clim. Change Res.* **2016**, *12*, 527–537. (In Chinese)
55. Huth, R.; Beck, C.; Philipp, A.; Demuzere, M.; Ustrnul, Z.; Cahynová, M.; Kyselý, J.; Tveito, O.E. Classifications of Atmospheric Circulation Patterns. *Ann. N. Y. Acad. Sci.* **2008**, *1146*, 105–152. [CrossRef] [PubMed]
56. Ning, G.; Yim, S.H.L.; Wang, S.; Duan, B.; Nie, C.; Yang, X.; Wang, J.; Shang, K. Synergistic effects of synoptic weather patterns and topography on air quality: A case of the Sichuan Basin of China. *Clim. Dyn.* **2019**, *53*, 6729–6744. [CrossRef]
57. Guo, M.; Zhang, M.; Wang, H.; Wang, L.; Liu, S.; Zong, L.; Zhang, Y.; Li, Y. Dual Effects of Synoptic Weather Patterns and Urbanization on Summer Diurnal Temperature Range in an Urban Agglomeration of East China. *Front. Environ. Sci.* **2021**, *9*, 672295. [CrossRef]
58. Philipp, A.; Beck, C.; Esteban, P.; Krennert, T.; Lochbihler, K.; Spyros, P.; Pianko-Kluczynska, K.; Post, P.; Alvarez, R.; Spekat, A.; et al. *Cost733 User Guide*; University of Augsburg: Augsburg, Germany, 2014.
59. Miao, Q.; Xu, X.; Pan, W. Winter Temperature Characteristics of Nanjing During 1951–2006. *J. Appl. Meteorol. Sci.* **2008**, *19*, 620–626. (In Chinese)
60. Frich, P.; Alexander, L.V.; Della-Marta, P.; Gleason, B.; Haylock, M.; Klein-Tank, A.M.G.; Peterson, T. Observed coherent changes in climatic extremes during the second half of the twentieth century. *Clim. Res.* **2002**, *19*, 193–212. [CrossRef]
61. Solomon, S.; Qin, D.; Manning, M.; Chen, Z.; Marquis, M.; Averyt, K.; Tignor, M.; Miller, H.L., Jr. (Eds.) *Climate Change 2007: The Physical Science Basis*; Cambridge University Press: Cambridge, UK, 2007; 996p.
62. Sun, B.; Wang, H.; Huang, Y.; Yin, Z.; Zhou, B.; Duan, M.K. Characteristics and causes of the hot-dry climate anomalies in China during summer of 2022. *Trans. Atmos. Sci.* **2022**, 1–9, *(online published)*. (In Chinese) [CrossRef]
63. Lu, R.; Xu, K.; Chen, R.; Chen, W.; Li, F.; Lv, C. Heat waves in summer 2022 and increasing concern regarding heat waves in general. *Atmos. Ocean. Sci. Lett.* **2023**, *16*, 100290. [CrossRef]
64. Chen, R.; Yin, P.; Wang, L.; Liu, C.; Niu, Y.; Wang, W.; Jiang, Y.; Liu, Y.; Liu, J.; Qi, J.; et al. Association between ambient temperature and mortality risk and burden: Time series study in 272 main Chinese cities. *BMJ* **2018**, *363*, k4306. [CrossRef]

Disclaimer/Publisher's Note: The statements, opinions and data contained in all publications are solely those of the individual author(s) and contributor(s) and not of MDPI and/or the editor(s). MDPI and/or the editor(s) disclaim responsibility for any injury to people or property resulting from any ideas, methods, instructions or products referred to in the content.

land

MDPI

Article

Combined Modification of Urbanization and Monsoon in Meiyu Precipitation Changes in the Megacity Shanghai, China

Ping Liang [1,2,*], Zhiqi Zhang [1], Wenjuan Huang [1], Qingfeng Zheng [1] and Yue Ma [3]

1 Key Laboratory of Cities' Mitigation and Adaptation to Climate Change in Shanghai, Shanghai Regional Climate Center, Shanghai 200030, China
2 Shanghai Typhoon Institute, China Meteorological Administration, Shanghai 200030, China
3 Shanghai Jiading District Meteorological Bureau, Shanghai 201800, China
* Correspondence: liangping1107@163.com

Abstract: The Meiyu season is a typical rainy season in East Asia that is controlled by summer monsoon. Despite extensive research on its impact, it is unclear how urbanization modifies precipitation during the Meiyu season in the background of the monsoon influence. To address this gap, this study investigated the effects of urbanization and monsoon on the modification of precipitation during the Meiyu season (PDM) in the megacity of Shanghai, China. Through homogenization analysis of the original observational data, we assessed the temporal and spatial variation in PDM in Shanghai during two stages of urbanization. Our findings revealed that both total precipitation and extreme daily precipitation during the Meiyu season in Shanghai have significantly increased since 1961. The spatial heterogeneity of PDM has also enhanced during the rapid urban process that has occurred since 1986. The long-term trend of increasing precipitation in Shanghai showed a synchronous variation with the East Asian subtropical summer monsoon (EASM) in 1961–2021. Over the interannual time scale, the significant positive correlation between PDM and EASM during the slow urbanization period (Stage 1: 1961–1985) changed to a non-significant correlation during the rapid urbanization period (Stage 2: 1986–2021), which was associated with the enhanced convective precipitation in Shanghai during the Meiyu season. Urbanization induced more convective precipitation and further weakened the association between PDM and EASM over the central city and nearby areas during Stage 2. The rapid urbanization process also resulted in increased differences in near-surface wind between urban and non-urban areas, which facilitated more PDM over the central city due to the urban friction effect and wind shear in Stage 2. Furthermore, our analysis suggests that the increase in precipitation may be associated with the enhanced coupling of cold air intrusion with the warmer climate background due to the urban heat effect occurring in Stage 2. These findings contribute to a better understanding of how urbanization and monsoons affect PDM in East Asian megacities and serve as a unique reference for climate prediction in this region.

Keywords: urbanization; Meiyu; monsoon; homogenized series; long-term change; interannual variation

1. Introduction

Meiyu is a major rainy season controlled by East Asian summer monsoon [1–3]. It usually refers to the persistent precipitation in June–July in the Yangtze River–Huaihe River Basin in China, South and Central Japan, and South Korea [4]. It is also called Baiu in Japan and Changma in South Korea [5,6]. Different from South Asian summer monsoon precipitation, the East Asian Meiyu is a product of the interaction between warm summer monsoon from low latitudes and cold air intrusions from middle to high latitudes [1,3,7–10]. Besides the natural variabilities associated with both the external forcing, including El Niño–Southern Oscillation (ENSO [11,12]), Atlantic Multidecadal Oscillation (AMO [13]), Pacific Decadal Oscillation (PDO [14]), and internal processes like Madden-Julian oscillation (MJO [3,15,16]) as well as North Atlantic Oscillation (NAO [17]), the precipitation during

East Asian Meiyu season (hereafter referring to June–July) is also impacted by global warming, urbanization, and aerosol effects [4].

With the expansion of urbanization across the globe, urban-rainfall modification has been reported in several studies [18–25]. Even though it is well known that urbanization can affect precipitation, studies varied based on the climate regime and the geographical locations of different cities [20,26,27]. In North China, the rapid urban expansion in Beijing was statistically correlated with summer rainfall reduction in the northeastern areas of the city since 1981 [20]. However, it had a positive effect on the changes in precipitation in autumn [28]. Obvious rain island effects can also be found in both the frequency of catastrophic storms above 100 mm [29] and the intensity of low-temperature rain and snow events in Beijing [30]. For eastern China, a larger precipitation frequency over urban areas took place in Nanjing and a significant enhancement of precipitation occurred in the downwind region of the city in the afternoon [31]. The spatial distribution of the storm frequency in Shanghai and its changing trend presented significant attributes of urban rain-island during its rapid urbanization process [21]. The heavy rainfall days also increased due to urbanization in Guangzhou [32]. The increase in precipitation mainly occurred downwind of the city belt in short-term rainstorm processes in the Yangtze River Delta [33]. Recent studies [24,27] have proved the consistency in spatial distributions of short-duration heavy precipitation with the heat island centers in megacities (Beijing, Shanghai) or city clusters [4,34–36]. However, the urban-rainfall modifications in different weather systems are not clear.

Some studies provided evidences that heavy rainfall over monsoon regions is a signature of urban-induced rainfall anomalies. For the Indian monsoon region, urban areas experienced fewer occurrences of light rainfall and significant higher occurrences of intense precipitation compared with nonurban regions during the monsoon season [37]. The rainfall islands parallel to the urban heat islands were also likely seen as coupled monsoon-urban induced effects under the weakened synoptic regime evidenced through monsoon low-level jet over Delhi, India [38]. In East Asian monsoon region, Meiyu is the major rainy season influenced by large-scale monsoon systems. However, the urban effect on rainfall during the Meiyu season is not clear and its related studies are rare. Using climate models, Ma and Zhang [39] studied the impact of urban expansion on large-scale precipitation by taking Meiyu as an example and showed decreased (increased) Meiyu precipitation over the Yangtze River (Huaihe River) basin. Quan et al. [40] suggested that precipitation increased significantly to the south of the Yangtze River due to large-scale urbanization-induced circulation changes. The results of model simulations are rather divergent and large uncertainty exists in assessing the urbanization effects due to differences in the choice of models. Shanghai is located in the lower reach of the Yangtze River Basin of China and is characterized by a distinct subtropic monsoon climate with Meiyu seasons. Shanghai is one of the largest megacities in China and has been experiencing rapid urbanization. It is interesting to evaluate the urbanization-monsoon modification in the PDM by taking the megacity Shanghai as a typical example.

In the present study, we examine the combined modifications of urbanization and monsoon in the change of PDM in Shanghai in 1961–2021. The study focuses on the following three topics: (a) the long-term changes in PDM associated with urbanization; (b) the monsoon influences associated with the long-term change of the PDM and (c) the urbanization effects on the changes in PDM during the two urbanization processes. Various observational and reanalysis products, as well as methods, are described in Section 2. The urbanization process in Shanghai and the homogenization test of PDM are presented in Section 3, and the impacts of monsoon and urbanization on variations in PDM are discussed in Sections 4 and 5, respectively. A summary and discussion are given in Section 6.

2. Materials and Methods

2.1. Data

Daily precipitation records were collected at Xujiahui (XJH) station from 1874 to 2021 and quality-controlled by Shanghai Meteorological Information Center (SMIC), Shanghai Meteorological Bureau. Daily observations, including precipitation, 2 m surface air temperature (SAT), and 10 m wind in 1961–2021, and the observational hourly precipitation in 1981–2021 were adopted from SMIC to investigate the spatial distribution of meteorological factors in Shanghai. The locations of the 11 basic stations in Shanghai are shown in Figure 1a. Related metadata of the observation history in Shanghai since 1874 and the Meiyu precipitation data in the Middle and Lower Reaches of the Yangtze River from 1885 to 2000 [41] were applied to check the homogeneity of the PDM series in Shanghai. The urbanization factors, including annual population density and paved road area in Shanghai from 1961 to 2021, were acquired from Shanghai Bureau of Statistics (https://tjj.sh.gov.cn/tjnj/index.html, accessed on 10 November 2022) to investigate the urbanization process.

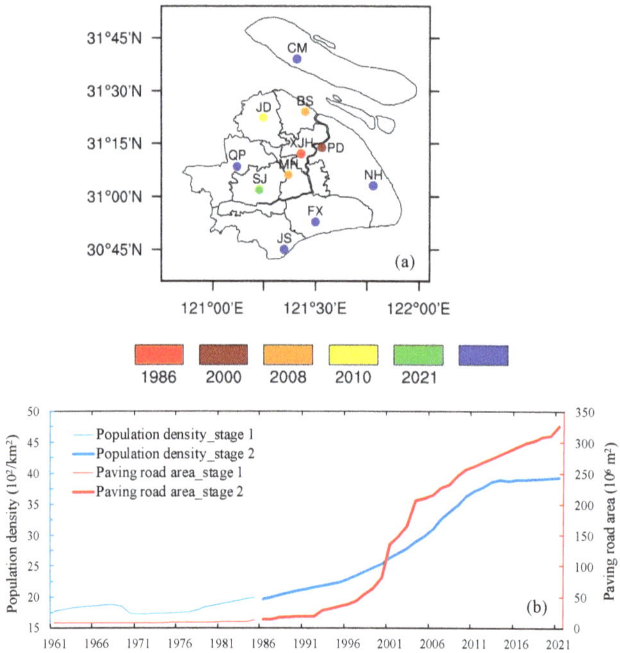

Figure 1. Locations of observational stations ((**a**), different colors denote the urbanized years) and evolution of the urbanization factors including population density (blue line) and paving road area (red line) (**b**) in Shanghai in 1961–2021.

Monthly atmospheric and precipitation reanalysis data from 1961 to 2021, with a 2.5° × 2.5° horizontal resolution, were downloaded from the National Center for Environmental Prediction (NCEP)/National Center for Atmospheric Research (NCAR) Reanalysis (NCEP/NCAR) [42] (https://psl.noaa.gov/data/gridded/data.ncep.reanalysis.html, accessed on 15 December 2022). NCEP reanalysis is not sensitive to urbanization or land-use effects [43] because surface observations over land are not used in the reanalysis [44]. Therefore, the precipitation and its associated atmospheric reanalysis from the NCEP/NCAR were adopted to investigate the impact of urbanization on Meiyu precipitation by comparing it with the surface observations.

2.2. Methods

Homogenization tests of precipitation datasets may be more reliable than the originals for urbanization impact analysis [45]. In the homogenization analysis of the PDM series, the extended version of Penalized Maximal F-test (PMF [46]), i.e., the "PMFred" algorithm [47] in the RHtestV5 software package was adopted to detect the possible change points without a reference series. Since Meiyu is a phenomenon influenced by the large-scale monsoon circulation, the Meiyu data in the Middle and Lower Reaches of the Yangtze River [41] was adopted as a reference series to check the homogeneity of Meiyu precipitation in Shanghai using the Penalized Maximal T-test (PMT) [48] in the RHtestV5 software package. All homogeneity tests in this study were analyzed at the 0.05 significance level.

Both impervious land cover [49,50] and population density were used to identify urban stations in order to decrease the uncertainty due to a single criterion. Following Liang and Ding [24] and Ma et al. [51], the urban stations were defined as having population densities above 3000 km^{-2} and an impervious fraction $\geq$0.5 in their surroundings, with a circular area of 10 km^2 (i.e., a radius of 1.785 km).

East Asian Subtropical Summer Monsoon Index (ESMI) was computed to investigate the impact of the summer monsoon on PDM. It was calculated as the anomaly of the difference in meridional moisture transport between South China and North China [52]. Positive (negative) ESMI corresponds to a strong (weak) subtropical summer monsoon. The ESMI had a significantly positive correlation with the summer rainfall over the middle and lower reaches of the Yangtze River, especially during the Meiyu season. According to Liang et al. [3], the correlation coefficient between the ESMI and PDM over the middle and lower reaches of the Yangtze River was 0.62 above the 0.01 significance level in the last 40 years (1979–2018).

Ensemble Empirical Mode Decomposition (EEMD [53]) was adopted to obtain the components of PDM over different time scales. EEMD is adaptive and derives optimal frequencies for decomposing data from the data itself, which provides a natural filter to separate components of different timescales [54]. The specific steps can be referred to Liang et al. [55].

Spatially normalized precipitation was used to investigate the spatial differences in the precipitation in Shanghai. It is calculated as follows [21]:

$$p_{sn,i} = (p_i - p_a)/\sigma_s, \quad (1)$$

where i denotes the number of the observational station; p_i and $p_{sn,i}$ are the original and the spatially normalized precipitation respectively; p_a and σ_s represent the spatial average and the standard deviation of precipitation at multi-stations respectively.

3. Urbanization Process in Shanghai and Precipitation Homogenization Test

3.1. Urbanization History in Shanghai

As a coastal megacity in the estuary of the Yangtze River Delta, Shanghai has experienced rapid urbanization since the 1980s. Figure 1a shows the start years when the different observational stations in Shanghai were identified to be urbanized based on population densities $\geq$ 3000 km^{-2} and impervious fractions $\geq$ 0.5. This suggests that urbanization has gradually expanded from the central city (XJH station) to the near suburbs since 1986. There were six urban stations and five nonurban stations in Shanghai in 2021. Pudong (PD) station was the second station urbanized in 2000, followed by Minhang (MH) and Baoshan (BS) in 2008, and Jiading (JD) and Songjiang (SJ) in 2010 and 2021, respectively. Therefore, the urbanization period in Shanghai since 1961 may be divided into the slow urbanization period (1961–1985, hereafter referred to as Stage 1) and the rapid urbanization period (1986–2021, hereafter referred to as Stage 2).

Urbanization is accompanied by increased population density, areas of built roads, etc. As can be seen in Figure 1b, both the population density and the area of paved road in Shanghai has increased steeply since the middle 1980s. The population density and the area of paved roads have been accelerated at $14.9 \times 10^2/km^2/year$ and $89.4 \times 10^6\ m^2/year$, respectively in Stage 2 (1986–2021), in sharp contrast to the slow urbanization period (Stage 1, i.e., 1961–1985).

3.2. Homogenization Test of Precipitation Series

The PMF test, without a reference series, was used to test homogenization in the PDM series at XJH since the records began in 1874. As shown in Figure 2a, no change point was identified in the precipitation series at XJH during the Meiyu season. Since Meiyu over the middle and lower reaches of the Yangtze River is a large-scale phenomenon influenced by the Asian monsoon system, Meiyu precipitation over the middle and lower reaches of the Yangtze River was also used as a reference series of Meiyu precipitation at XJH. No change point of the Meiyu precipitation series at XJH was identified based on the PMT test by using the reference series (Figure 2b). Therefore, it can be inferred that the precipitation series at XJH during the Meiyu season may be homogeneous based on a century scale. In addition to XJH, similar results were obtained from homogenization tests carried out on the observational precipitation series data collected at the other 10 stations since 1961. By taking the average precipitation series of the 11 stations in Shanghai as an example, no change point was identified based on the PMF test (Figure 2c). Therefore, the original precipitation observations in Shanghai during the Meiyu season can be viewed as homogeneous.

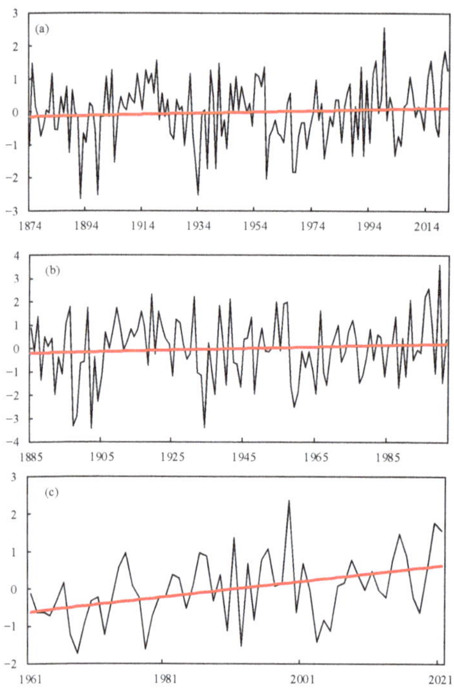

Figure 2. Homogenization test of precipitation series: (**a**) anomalies of normal transformed PDM at XJH in 1874–2021 (Red line denotes the regression fit using the PMF test); (**b**) anomalies of normal transformed Meiyu precipitation series at XJH in 1885–2001 with reference series of Meiyu precipitation over the Yangtze River (Red line denotes the regression fit using the PMT test); (**c**) anomalies of normal transformed average PDM of 11 stations in Shanghai during the 1961–2021 period (Red line denotes the regression fit using the PMF test).

4. Long-Term Changes in PDM Associated with Urbanization

Figure 3 shows the variation in total precipitation and extreme daily precipitation at XJH during the Meiyu seasons in 1874–2021. The annual extreme daily precipitation is defined as the average of the 5% heaviest daily records during the Meiyu season in every year. It can be clearly seen that both total precipitation and extreme daily precipitation exhibited significant (above 0.01 significance level) increasing trends from 1961, while no distinct trends were observed in 1874–1960. The trends of increase in both the total precipitation and extreme daily precipitation at XJH since 1961 are consistent with the trend of increased warming at XJH [55], which is associated with the increase in water vapor based on the Clausius-Clapeyron Equation [56].

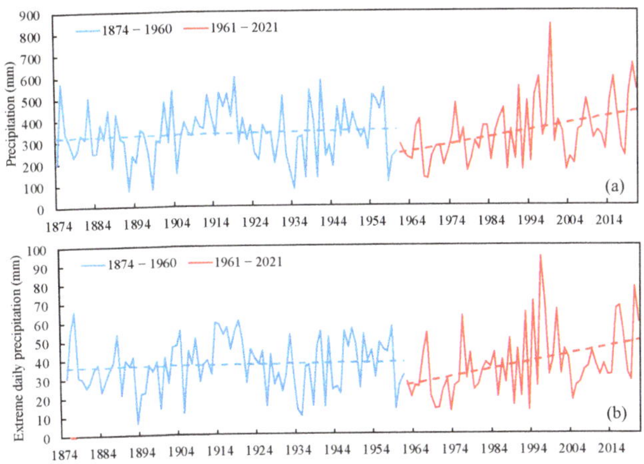

Figure 3. Variation in the total precipitation (**a**) and extreme daily precipitation (**b**) at XJH during the Meiyu seasons of 1874–2021 (Dashed lines are the corresponding linear trends).

In order to investigate the spatial differences in the Meiyu rainfall in Shanghai, Figure 4a,b shows the average PDM during the two urbanization stages in Shanghai. Compared with Stage 1 (the slow urban process, Figure 4a), the increase in PDM varies from 71 mm to 111.8 mm in Shanghai during the rapid urban process (Stage 2, Figure 4b). Meanwhile, the spatial distribution of the PDM transformed from the south-north mode in Stage 1 to the distinct urban island mode in Stage 2. The spatial standard deviation of PDM exhibited a trend of significant (above 0.01 significance level) increase accompanying the increase in PDM in Shanghai (Figure 4c). On average, the spatial standard deviation increased by about 52% in Stage 2 compared with Stage 1. In other words, spatial heterogeneity in PDM increased with the strengthening of PDM during the rapid urban process, with more PDM concentrated in the central urban city.

Further analysis found that the increase in strong convection events (i.e., hourly precipitation greater than 20 mm) contributed to the PDM during Stage 2. As shown in Figure 5a,b, two strong convection events per year usually occurred in Stage 2, with an increase of 50% compared with that in Stage 1. For the spatial distribution (Figure 5c,d), the large value regions of both the frequency and the normalized precipitation of strong convection events transferred from the southeast part of Shanghai in Stage 1 to the central city including XJH and PD stations in Stage 2. In other words, the strong convections during the Meiyu season also exhibited distinct characteristics of urban rain island distribution during the rapid urbanization process.

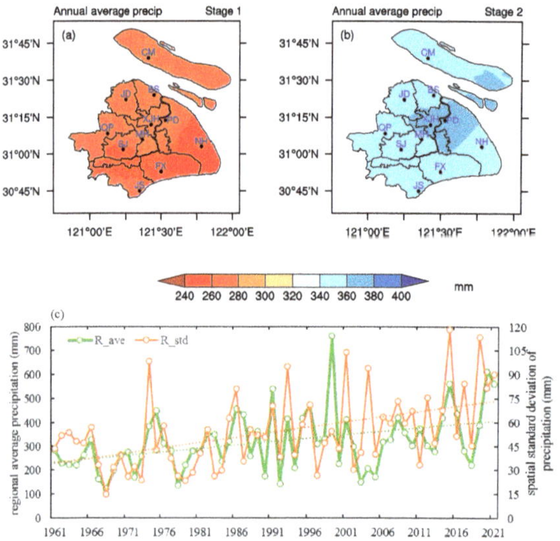

Figure 4. Spatial distribution of the average PDM in Shanghai during Stages 1/2 (**a**,**b**); i.e., the slow/fast urban process) and the spatial standard deviation of the PDM in Shanghai in 1961–2021 ((**c**), yellow and green curves denote the PDM; yellow/green lines represent linear trends).

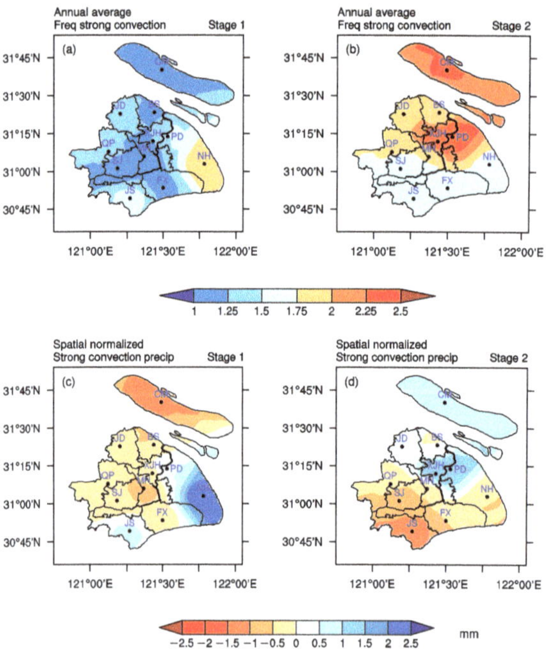

Figure 5. Spatial distribution of the frequency (**a**,**b**) and the spatially normalized precipitation (**c**,**d**) of strong convection events (hourly precipitation greater than 20 mm) in Shanghai during the Meiyu season during the slow/fast (**a**–**d**) urban processes.

5. Influence of Monsoon on PDM during Two Different Urbanization Processes

Meiyu over the Yangtze River, including in Shanghai, is strongly influenced by the East Asian subtropical summer monsoon (EASM). What is the impact of the EASM on the PDM in accompanying the development of urbanization in Shanghai? EEMD was adopted to isolate variabilities in the PDM and EASM index (EASMI) at different timescales during the Meiyu seasons in 1961–2021. The long-term trend and interannual components are shown in Figure 6a,b respectively. It can be clearly seen that the long-term trend component of the PDM has a variation that is synchronous with that of the EASM, with their correlation coefficient as high as 0.91 (Figure 6a). Note that the trend of EEMD is nonlinear [53]. The trend of decreasing (increasing) PDM in Shanghai occurred before (after) the mid-1980s, and the amplitude of the trend varied over different periods. The case is similar to the variation in the trend of the EASM. The PDM in Shanghai showed stable increasing trend with the increase of EASM in 1961–2021. It reflects the regional influence of the EASM on PDM at the long-term scale. For the interannual variability, a distinct change took place in the relationship between the PDM in Shanghai and EASM. As shown in Figure 6b, the significant positive correlation (correlation coefficient 0.59 above 0.01 significance level) between the PDM and EASM during Stage 1 changed to a nonsignificant correlation (correlation coefficient 0.24) during Stage 2. This suggests that the relationship between EASM and PDM at the interannual time scale distinctively weakened during the rapid urbanization process.

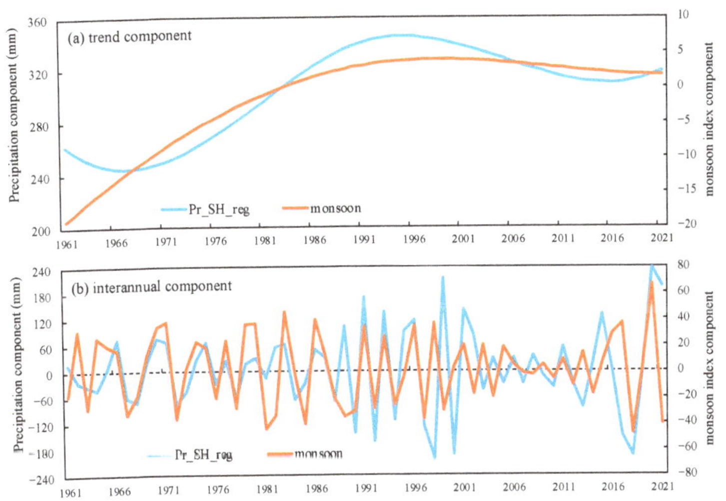

Figure 6. The long-term trend (**a**) and the interannual (**b**) components of the regional average PDM in Shanghai (blue curves) together with their corresponding EASMI components (orange curves) in 1961–2021 based on EEMD analysis.

The interannual component of the reanalyzed precipitation in Shanghai was also adopted to investigate the relationship between two kinds of precipitation (large-scale precipitation and convective precipitation) in PMD with EASM. As shown in Table 1, in Stage 1, both total precipitation during the Meiyu season and its large-scale precipitation component had a significant positive correlation with EASM, while the convective precipitation was not correlated with EASM. The contrary was true in Stage 2, i.e., both total precipitation and the large-scale precipitation component had no significant correlation with EASM. However, the convective precipitation component had a significant relationship with EASM in Stage 2. In other words, in the background of rapid urbanization, the PMD associated with EASM exhibited more convective properties. It may be associated with enhanced atmospheric instability and moisture transportation over Shanghai [24].

The conversion of the precipitation property from large-scale to convective may contribute to the weakening of the relationship between PDM and EASM at the interannual scale.

Table 1. Correlation coefficients between interannual components of precipitation reanalysis and East Asian summer monsoon index during Meiyu season (bold numbers denote above 0.05 significance level).

Precipitation Component	Stage 1 (1961–1985)	Stage 2 (1986–2021)
Total precipitation	**0.38**	0.07
Large-scale precipitation	**0.49**	−0.08
Convective precipitation	0.07	**0.32**

The spatial distributions of the correlation coefficients between the PDM and EASM during the two stages of urbanization are shown in Figure 7. During Stage 1, with no urban stations, the correlations between PDM and EASM were significant at all stations in Shanghai. However, the significant correlation area was restricted to the northern part (JD, BS, and CM stations) and the southern part (FX station) of Shanghai in Stage 2. The relationship between the PDM and EASM over the central city and nearby suburbs (XJH, PD, MH, and SJ stations) distinctively weakened during Stage 2. This suggests that urbanization effects further weakened the association between PDM and EASM, in addition to the above-mentioned influence of regional convective rainfall at the inter-annual scale.

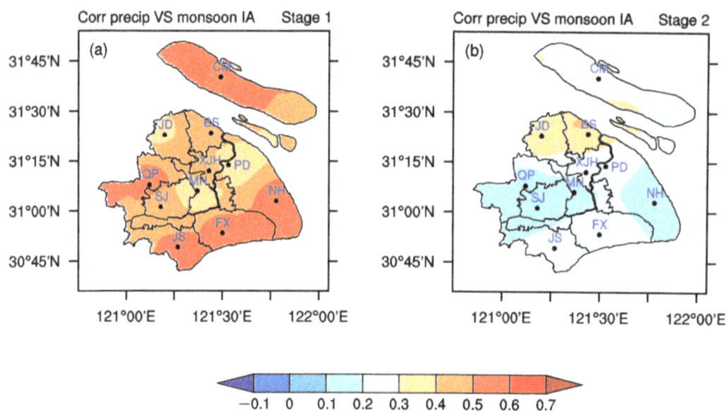

Figure 7. Spatial distribution of the correlation coefficients between the PDM and EASM during the slow (**a**) and rapid (**b**) urbanization process.

6. Effects of Urbanization on PDM in Two Different Urbanization Processes

The urban heat island is a well-known urbanization effect. How does the urban heat effect impact PDM? Figure 8 shows the average 2 m surface air temperature (SAT) during the Meiyu season, and its correlation coefficients with the PDM in Shanghai during the rapid urbanization process (Stage 2). Similar to Stage 1 (figure omitted), the PDM is negatively correlated with the SAT during the Meiyu season when the Meiyu usually occurs under the persistent interaction of cold air intrusion and warm air transported by the Asian monsoon. Meanwhile, the coupling of SAT and PDM was strengthened in Stage 2, with an average correlation coefficient of −0.48 between them. The coupling of temperature and PDM further increased over the urban area. As shown in Figure 8, the negative correlations were most significant in the central urban areas (XJH). For the spatial difference in Shanghai, the negative coupling between SAT and PDM was significantly correlated with the average SAT in Shanghai, with a correlation coefficient of −0.67 above the 0.01 significance level). In other words, over the stations with warmer climate backgrounds impacted by the urban heat effect, more PDM occurred when cold air intrusion interacted with the monsoon. This may be associated with more initial water vapor stored in the warmer environment.

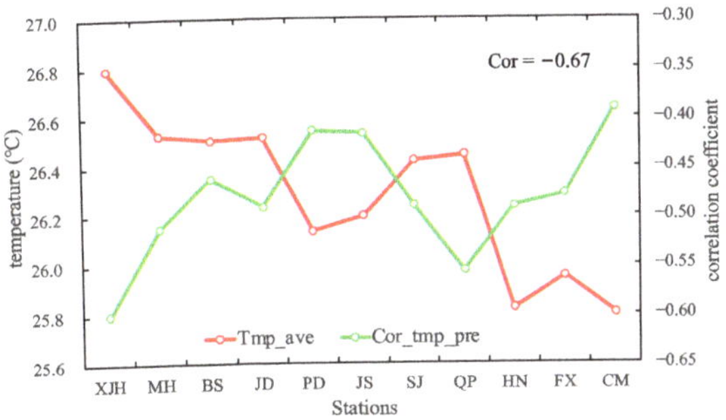

Figure 8. The average SAT during the Meiyu season (red curve) and its correlation coefficients (green curve) with PDM at the 11 observational stations (marked on the x-axis) in Stage 2 (The number on the upper right corner denotes the correlation coefficient between the two series).

Urban friction effect also exerts an impact on precipitation during the Meiyu season. Figure 9a,b shows the spatial distribution of the 10 m wind speed in Shanghai during the Meiyu season in Stage 1 and 2 respectively. It can be clearly seen that the wind speeds were generally decreased in Shanghai during the Meiyu season under the background of fast urbanization compared with Stage 1. Thereinto, the wind speeds were decreased by 1.3 m/s (accounting for 40% of that in Stage 1) at XJH and PD stations over the central city in Stage 2. The amplitude of the weakening of the wind speed was the smallest at the nonurban JS station over the southern suburb of Shanghai. Meanwhile, the urban weak wind phenomenon was obvious over the central city and nearby area in Stage 2, while the relatively weak wind was mainly located over the downwind direction of Shanghai in Stage 1. The urban weak wind distribution was consistent with the urban rainfall island phenomena in both the total PDM (Figure 4b) and its strong convection components (Figure 5b,d). Further analysis shows that the correlation coefficient of the spatial normalized PDM and the wind speed during the Meiyu season in Shanghai was −0.55 and −0.62 in Stage 1 and Stage 2, respectively. This means that the smaller near-surface wind speed, the more PDM in both Stages 1 and 2. Additionally, the rapid urbanization in Stage 2 can contribute to the increased PDM due to the urban friction effect and its related convergence condition.

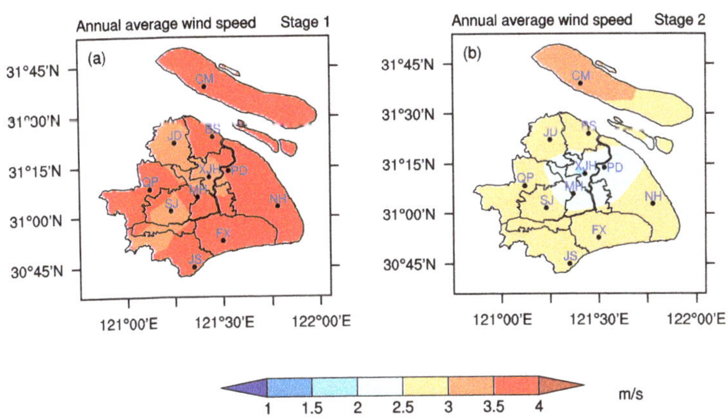

Figure 9. Spatial distribution of the average 10m wind speed in Shanghai during the slow (**a**) and rapid (**b**) urbanization processes.

Figure 10 further shows the wind-rose diagrams of the typical urban (XJH) and nonurban (JS) stations during the two stages. There was little change in the near-surface wind at JS station (nonurban station) during the Meiyu seasons from Stage 1 to Stage 2

(Figure 10b). However, there was an obvious change observed at XJH station with the change from the unurbanized process to the urbanized process. As can be seen in Figure 10a, more quiet wind occurred, and less prevalent wind took place during the urbanized process. By combing Figures 9 and 10, the enhanced difference in both wind speed and direction between the urban station and the outskirt nonurban areas was favorable for the wind shear and its associated rainfall over the central city of Shanghai during the Meiyu season in the rapid urbanization stage.

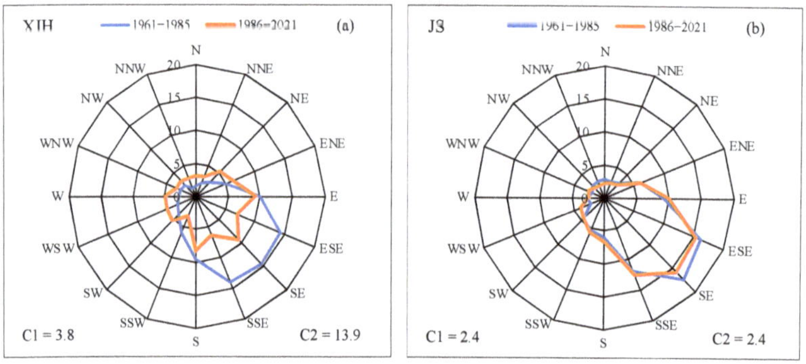

Figure 10. Wind-rose diagram at XJH (**a**) and JS (**b**) stations during Stages 1 (blue curve) and 2 (orange curve). C1 and C2 denote the frequencies of quiet wind during Stages 1 and 2.

7. Discussion and Summary

Based on the homogenization test of precipitation in Shanghai, we evaluated the combined impact of urbanization and monsoon on precipitation in Shanghai during the Meiyu season (PDM) by analyzing two different urbanization processes in 1961–2021.

Using the homogenization tests, we confirmed the observational PDM series at the XJH station (over a century) and 10 other stations (since 1961) to be homogeneous. For the long-term change, both total precipitation and extreme daily precipitation during the Meiyu season have significantly increased in Shanghai since 1961. During the rapid urbanization process since 1986, the spatial heterogeneity of PDM has increased with the strengthening of PDM. More PDM is concentrated in the central urban city. Additionally, the strong convection occurring during the Meiyu season has also exhibited an urban rain island distribution throughout the rapid urbanization process.

The long-term increasing trend of PDM shows synchronization with that of the EASM in 1961–2021, demonstrating the influence of EASM on PDM at a regional scale. However, at the interannual time scale, the significant positive correlation between PDM and EASM has changed to a nonsignificant correlation from the slow urbanization process (Stage 1) to the rapid urbanization process (Stage 2). More convective precipitation may contribute to the weakening of the relationship at the interannual time scale.

In the background of rapid urbanization, the urbanization effect may further weaken the association between PDM and EASM over the central city and nearby areas in Shanghai by inducing more convective precipitation. Furthermore, impacted by the urban heat effect in Stage 2, more PDM over the central city and nearby areas in Shanghai is associated with the enhanced coupling of cold air intrusion with the warmer climate background. In addition, the rapid urbanization enhanced the difference in the near-surface wind between the urban station and the outskirt nonurban areas, which facilitated more PDM over the central city of Shanghai by way of increasing wind shear.

As pointed out by Liang et al. [55], a notable inhomogeneous change point in SAT at XJH station occurred in 1954 due to the change of observation schedule in 1954. The observations of precipitation at XJH changed from four times at 1, 7, 13, and 19 o'clock (Beijing time, also hereinafte) per day in 1954–1960 to 2, 8, 14, and 20 o'clock per day after 1960. The homogenization of the daily precipitation record may be influenced by the change of the observation schedule. Due to lack of observations of hourly rainfall before 1980, two random resampling based on the hourly precipitation observed since 1981 were calculated according to the above two different daily observation schedules. It was found that the extreme daily precipitation may be influenced by the observation schedules. By using the observations at 1, 7, 13, and 19 o'clock per day, the extreme daily precipitation defined by the 99.5% percentile during the Meiyu season may be generally underestimated by about 0.8 mm. Considering the small amplitude in the extreme daily precipitation deviation caused by the change of the observation schedule, the long-term change of the extreme daily precipitation in this paper is reasonable.

Besides the thermal and dynamic impact of urbanization on the PDM described in this study, the aerosol effects associated with urbanization may also exert influences on the change in PDM. Jung et al. [57] pointed out that aerosols exerted significant indirect effects in mid-and low-level clouds, resulting in an increase in the cloud particle radius and enhanced precipitation intensity during the Meiyu period in the Yangtze-Huaihe Basin. It is necessary to further study the mechanism of the abnormal activities of Meiyu from the perspective of multi-factor interactions at the multi-scale, including atmospheric chemistry, urbanization, monsoons, and global warming.

Author Contributions: Conceptualization, P.L.; methodology, P.L.; investigation, P.L., Z.Z., W.H., Q.Z. and Y.M.; data curation, W.H. and Q.Z.; writing—original draft preparation, P.L.; writing—review and editing, P.L. and Z.Z. All authors have read and agreed to the published version of the manuscript.

Funding: This research was jointly supported by the National Natural Science Foundation of China (Grant 42175056), the Natural Science Foundation of Shanghai (21ZR1457600), the Shanghai Sailing Program (23YF1440100), the China Meteorological Administration Innovation and Development Project (CXFZ2022J009), the China Meteorological Administration Review Summary Project (FPZJ2023-044), the Open Fund of the Key Laboratory of Cities' Mitigation and Adaptation to Climate Change in Shanghai, China Meteorological Administration, and the Key Innovation Teams of the China Meteorological Administration (CMA2023ZD03).

Data Availability Statement: Precipitation, 2 m surface air temperature, and 10 m wind observations in Shanghai were obtained from the Shanghai Meteorological Information Center (SMIC), Shanghai Meteorological Bureau. The urbanization factors in Shanghai were obtained from the Shanghai Bureau of Statistics (https://tjj.sh.gov.cn/tjnj/index.html (accessed on 10 November 2022)). Monthly atmospheric and precipitation reanalysis data were downloaded from the National Centers for Environmental Prediction (NCEP)-National Center for Atmospheric Research (NCAR) Reanalysis (NCEP/NCAR) (https://psl.noaa.gov/data/gridded/data.ncep.reanalysis.html (accessed on 15 December 2022)).

Acknowledgments: We thank Zongwei Yan for his thoughtful advice on the homogenization tests, and the reviewers for their suggestions.

Conflicts of Interest: The authors declare no conflict of interest.

References

1. Tao, S.-Y.; Chen, L.-X. A Review of Recent Research on the East Asian Summer Monsoon in China. In *Monsoon Meteorology*; Chang, C.P., Krishnamurti, T.N., Eds.; Oxford University Press: Oxford, UK, 1987; pp. 60–92.
2. Ding, Y. *Monsoons over China*; Atmospheric and Oceanographic Sciences Library; Springer: Dordrecht, The Netherlands, 1994; Volume 16, ISBN 978-94-015-8302-2.
3. Liang, P.; Hu, Z.-Z.; Ding, Y.; Qian, Q. The Extreme Mei-Yu Season in 2020: Role of the Madden-Julian Oscillation and the Cooperative Influence of the Pacific and Indian Oceans. *Adv. Atmos. Sci.* **2021**, *38*, 2040–2054. [CrossRef]
4. Ding, Y.; Liang, P.; Liu, Y.; Zhang, Y. Multiscale Variability of Meiyu and Its Prediction: A New Review. *J. Geophys. Res. Atmos.* **2020**, *125*, e2019JD031496. [CrossRef]
5. Ninomiya, K.; Muraki, H. Large-Scale Circulations over East Asia during Baiu Period of 1979. *J. Meteorol. Soc. Jpn.* **1986**, *64*, 409–429. [CrossRef]
6. Oh, J.-H.; Kwon, W.-T.; Ryoo, S.-B. Review of the Researches on Changma and Future Observational Study (Kormex). *Adv. Atmos. Sci.* **1997**, *14*, 207–222. [CrossRef]

7. Ding, Y. Summer Monsoon Rainfalls in China. *J. Meteorol. Soc. Jpn.* **1992**, *70*, 373–396. [CrossRef]
8. Ding, Y. Seasonal March of the East Asian Monsoon. In *East Asian Monsoon*; World Scientific Series on Asia-Pacific Weather and Climate; World Scientific: Singapore, 2004; Volume 2, pp. 3–53, ISBN 978-981-238-769-1.
9. Lau, K.M.; Ding, Y.; Wang, J.-T.; Johnson, R.; Keenan, T.; Cifelli, R.; Gerlach, J.; Thiele, O.; Rickenbach, T.; Tsay, S.-C.; et al. A Report of the Field Operations and Early Results of the South China Sea Monsoon Experiment (SCSMEX). *Bull. Amer. Meteor. Soc.* **2000**, *81*, 1261–1270. [CrossRef]
10. Ding, Y.; Liu, Y.; Hu, Z.-Z. The Record-Breaking Mei-Yu in 2020 and Associated Atmospheric Circulation and Tropical SST Anomalies. *Adv. Atmos. Sci.* **2021**, *38*, 1980–1993. [CrossRef]
11. Bjerknes, J. A Possible Response of the Atmospheric Hadley Circulation to Equatorial Anomalies of Ocean Temperature. *Tellus A Dyn. Meteorol. Oceanogr.* **1966**, *18*, 820–829. [CrossRef]
12. Liang, P.; Hu, Z.Z.; Liu, Y.; Yuan, X.; Li, X.; Jiang, X. Challenges in predicting and simulating summer rainfall in the eastern China. *Clim. Dyn.* **2019**, *52*, 2217–2233. [CrossRef]
13. Enfield, D.B.; Mestas-Nuñez, A.M.; Trimble, P.J. The Atlantic Multidecadal Oscillation and Its Relation to Rainfall and River Flows in the Continental U.S. *Geophys. Res. Lett.* **2001**, *28*, 2077–2080. [CrossRef]
14. Mantua, N.J.; Hare, S.R.; Zhang, Y.; Wallace, J.M.; Francis, R.C. Pacific Interdecadal Climate Oscillation with Impacts on Salmon Production. *Am. Meteorol. Soc.* **1997**, *78*, 1069–1079. [CrossRef]
15. Liang, P.; Lin, H. Sub-seasonal prediction over East Asia during boreal summer using the ECCC monthly forecasting system. *Clim. Dyn.* **2018**, *52*, 1007–1022. [CrossRef]
16. Zhu, Z.; Zhou, Y.; Jiang, W.; Fu, S.; Hsu, P. Influence of compound zonal displacements of the South Asia high and the western Pacific. subtropical high on Meiyu intraseasonal variation. *Clim. Dyn.* **2023**, 1–17. [CrossRef]
17. Yan, Y.; Liu, B.; Zhu, C.; Lu, R.; Jiang, N.; Ma, S. Subseasonal forecast barrier of the North Atlantic oscillation in S2S models during the extreme mei-yu rainfall event in 2020. *Clim. Dyn.* **2022**, *58*, 2913–2925. [CrossRef]
18. Chen, L.-X.; Zhu, W.; Zhou, X.; Zhou, Z. Characteristics of the Heat Island Effect in Shanghai and Its Possible Mechanism. *Adv. Atmos. Sci.* **2003**, *20*, 991–1001. [CrossRef]
19. Wang, X.; Wang, Z.; Qi, Y.; Guo, H. Effect of Urbanization on the Winter Precipitation Distribution in Beijing Area. *Sci. China Ser. D Earth Sci.* **2009**, *52*, 250–256. [CrossRef]
20. Zhang, C.L.; Chen, F.; Miao, S.G.; Li, Q.C.; Xia, X.A.; Xuan, C.Y. Impacts of Urban Expansion and Future Green Planting on Summer Precipitation in the Beijing Metropolitan Area. *J. Geophys. Res.* **2009**, *114*, D02116. [CrossRef]
21. Liang, P.; Ding, Y.; He, J.-H.; Tang, X. Study of Relationship between Urbanization Speed and. Change in Spatial Distribution of Rainfall over Shanghai. *J. Trop. Meteorol.* **2013**, *19*, 475–483.
22. Yang, L.; Tian, F.; Smith, J.A.; Hu, H. Urban Signatures in the Spatial Clustering of Summer Heavy Rainfall Events over the Beijing Metropolitan Region: Urban Modification of Heavy Rainfall. *J. Geophys. Res. Atmos.* **2014**, *119*, 1203–1217. [CrossRef]
23. Zhong, S.; Yang, X.-Q. Ensemble Simulations of the Urban Effect on a Summer Rainfall Event in the Great Beijing Metropolitan Area. *Atmos. Res.* **2015**, *153*, 318–334. [CrossRef]
24. Liang, P.; Ding, Y. The Long-Term Variation of Extreme Heavy Precipitation and Its Link to Urbanization Effects in Shanghai during 1916–2014. *Adv. Atmos. Sci.* **2017**, *34*, 321–334. [CrossRef]
25. Liu, J.; Niyogi, D. Meta-Analysis of Urbanization Impact on Rainfall Modification. *Sci Rep* **2019**, *9*, 7301. [CrossRef]
26. Dou, J.; Wang, Y.; Bornstein, R.; Miao, S. Observed Spatial Characteristics of Beijing Urban Climate Impacts on Summer Thunderstorms. *J. Appl. Meteorol. Climatol.* **2015**, *54*, 94–105. [CrossRef]
27. Yang, P.; Ren, G.; Yan, P. Evidence for a Strong Association of Short-Duration Intense Rainfall with Urbanization in the Beijing Urban Area. *J. Clim.* **2017**, *30*, 5851–5870. [CrossRef]
28. Song, X.; Mo, Y.; Xuan, Y.; Wang, Q.J.; Wu, W.; Zhang, J.; Zou, X. Impacts of Urbanization on Precipitation Patterns in the Greater Beijing–Tianjin–Hebei Metropolitan Region in Northern China. *Environ. Res. Lett.* **2021**, *16*, 014042. [CrossRef]
29. Hu, H. Spatiotemporal Characteristics of Rainstorm-Induced Hazards Modified by Urbanization in Beijing. *J. Appl. Meteorol. Climatol.* **2015**, *54*, 1496–1509. [CrossRef]
30. Han, Z.; Yan, Z.; Li, Z.; Liu, W.; Wang, Y. Impact of Urbanization on Low-Temperature Precipitation in Beijing during 1960–2008. *Adv. Atmos. Sci.* **2014**, *31*, 48–56. [CrossRef]
31. Yang, B.; Zhang, Y.; Qian, Y. Simulation of Urban Climate with High-Resolution WRF Model: A Case Study in Nanjing, China. *Asia-Pac. J. Atmos. Sci.* **2012**, *48*, 227–241. [CrossRef]
32. Liao, J.; Wang, X.; Li, Y.; Xia, B. An Analysis Study of the Impacts of Urbanization on Precipitation in Guangzhou. *J. Meteorol. Sci.* **2011**, *31*, 384–390. [CrossRef]
33. Wan, H.; Zhong, Z.; Yang, X.; Li, X. Impact of City Belt in Yangtze River Delta in China on a Precipitation Process in Summer: A Case Study. *Atmos. Res.* **2013**, *125–126*, 63–75. [CrossRef]
34. Wang, J.; Feng, J.; Yan, Z.; Hu, Y.; Jia, G. Nested High-Resolution Modeling of the Impact of Urbanization on Regional Climate in Three Vast Urban Agglomerations in China: Climate impact of urbanization in China. *J. Geophys. Res.* **2012**, *117*, D21103. [CrossRef]
35. Feng, J.; Wang, J.; Yan, Z. Impact of Anthropogenic Heat Release on Regional Climate in Three Vast Urban Agglomerations in China. *Adv. Atmos. Sci.* **2014**, *31*, 363–373. [CrossRef]

36. Jiang, X.; Luo, Y.; Zhang, D.-L.; Wu, M. Urbanization Enhanced Summertime Extreme Hourly Precipitation over the Yangtze River Delta. *J. Clim.* **2020**, *33*, 5809–5826. [CrossRef]
37. Kishtawal, C.M.; Niyogi, D.; Tewari, M.; Pielke, R.A.; Shepherd, J.M. Urbanization Signature in the Observed Heavy Rainfall Climatology over India. *Int. J. Climatol.* **2010**, *30*, 1908–1916. [CrossRef]
38. Chalakkal, J.B.; Mohan, M. Is the Monsoon Climatology Observed over the National Capital Region, Delhi Indicative of an Urban-Monsoon Linkage on Rainfall Modification? *Urban Clim.* **2022**, *46*, 101289. [CrossRef]
39. Ma, X.; Zhang, Y. Numerical Study of the Impacts of Urban Expansion on Meiyu Precipitation over Eastern China. *J. Meteorol. Res.* **2015**, *29*, 237–256. [CrossRef]
40. Quan, J.; Xue, Y.; Duan, Q.; Liu, Z.; Oleson, K.W.; Liu, Y. Numerical Investigation and Uncertainty Analysis of Eastern China's Large-Scale Urbanization Effect on Regional Climate. *J. Meteorol. Res.* **2021**, *35*, 1023–1040. [CrossRef]
41. Xu, Q.; Yang, Y.-W.; Yang, Q.-M. The Meiyu in Middle-Lower Reaches of Yangtze River during 116 Recent Years (I). *Torrential Rain Disaster* **2001**, *1*, 44–53.
42. Kalnay, E.; Kanamitsu, M.; Kistler, R.; Collins, W.; Deaven, D.; Gandin, L.; Iredell, M.; Saha, S.; White, G.; Woollen, J.; et al. The NCEP/NCAR 40-Year Reanalysis Project. *Bull. Am. Meteorol. Soc.* **1996**, *77*, 437–471. [CrossRef]
43. Kalnay, E.; Cai, M. Impact of urbanization and landuse change on climate. *Nature* **2003**, *423*, 528–531. [CrossRef]
44. Kistler, R.; Kalnay, E.; Collins, W.; Saha, S.; White, G.; Woollen, J.; Chelliah, M.; Ebisuzaki, W.; Kanamitsu, M.; Kousky, V.; et al. The NCEP/NCAR 50-year reanalysis: Monthly means CD-ROM and documentation. *Bull. Am. Meteorol. Soc.* **2000**, *82*, 247–267. [CrossRef]
45. Li, Z.; Yan, Z.; Tu, K.; Wu, H. Changes of Precipitation and Extremes and the Possible Effect of Urbanization in the Beijing Metropolitan Region during 1960–2012 Based on Homogenized Observations. *Adv. Atmos. Sci.* **2015**, *32*, 1173–1185. [CrossRef]
46. Wang, X.L. Penalized Maximal F Test for Detecting Undocumented Mean Shift without Trend Change. *J. Atmos. Ocean. Technol.* **2008**, *25*, 368–384. [CrossRef]
47. Wang, X.L.; Chen, H.; Wu, Y.; Feng, Y.; Pu, Q. New Techniques for the Detection and Adjustment of Shifts in Daily Precipitation Data Series. *J. Appl. Meteorol. Climatol.* **2010**, *49*, 2416–2436. [CrossRef]
48. Wang, X.L.; Wen, Q.H.; Wu, Y. Penalized Maximal t Test for Detecting Undocumented Mean Change in Climate Data Series. *J. Appl. Meteorol. Climatol.* **2007**, *46*, 916–931. [CrossRef]
49. Gong, P.; Liu, H.; Zhang, M.; Li, C.; Wang, J.; Huang, H.; Clinton, N.; Ji, L.; Li, W.; Bai, Y.; et al. Stable Classification with Limited Sample: Transferring a 30-m Resolution Sample Set Collected in 2015 to Mapping 10-m Resolution Global Land Cover in 2017. *Sci. Bull.* **2019**, *64*, 370–373. [CrossRef] [PubMed]
50. Gong, P.; Li, X.; Zhang, W. 40-Year (1978–2017) Human Settlement Changes in China Reflected by Impervious Surfaces from Satellite Remote Sensing. *Sci. Bull.* **2019**, *64*, 756–763. [CrossRef]
51. Ma, Y.; Liang, P.; Grimmond, S.; Yang, X.; Lyu, J.; Ding, Y. Three-Dimensional Urban Thermal Effect across a Large City Cluster during an Extreme Heat Wave: Observational Analysis. *J. Meteorol. Res.* **2022**, *36*, 387–400. [CrossRef]
52. Liang, P.; Tang, X.; He, J.-H.; Chen, L.-X. An East Asian Subtropical Summer Monsoon Index Defined by Moisture Transport. *J. Trop. Meteorol.* **2008**, *14*, 61–64.
53. Wu, Z.; Huang, N.E. Ensemble Empirical Mode Decomposition: A Noise-Assisted Data Analysis Method. *Adv. Adapt. Data Anal.* **2009**, *1*, 1–41. [CrossRef]
54. Huang, N.E.; Wu, Z. A Review on Hilbert-Huang Transform: Method and Its Applications. *Rev. Geophys.* **2008**, *46*, 1–23. [CrossRef]
55. Liang, P.; Yan, Z.-W.; Li, Z. Climatic Warming in Shanghai during 1873–2019 Based on Homogenised Temperature Records. *Adv. Clim. Chang. Res.* **2022**, *13*, 496–506. [CrossRef]
56. Held, I.M.; Soden, B.J. Robust Responses of the Hydrological Cycle to Global Warming. *J. Clim.* **2006**, *19*, 5686–5699. [CrossRef]
57. Jung, W.-S.; Panicker, A.S.; Lee, D.-I.; Park, S.-H. Estimates of Aerosol Indirect Effect from Terra MODIS over Republic of Korea. *Adv. Meteorol.* **2013**, *2013*, 976813. [CrossRef]

Disclaimer/Publisher's Note: The statements, opinions and data contained in all publications are solely those of the individual author(s) and contributor(s) and not of MDPI and/or the editor(s). MDPI and/or the editor(s) disclaim responsibility for any injury to people or property resulting from any ideas, methods, instructions or products referred to in the content.

land

Article

Urbanization-Induced Diurnal Variation in Short-Duration Rainfall Events in Wuhan, China

Yanlin Mao [1,2], Guoyu Ren [1,3] and Suonam Kealdrup Tysa [4,*]

1 Department of Atmospheric Science, School of Environmental Studies, China University of Geosciences, Wuhan 430070, China
2 CMA-NJU Joint Laboratory for Climate Prediction Studies, School of Atmospheric Sciences, Nanjing University, Nanjing 210023, China
3 Laboratory for Climate Studies, National Climate Center, China Meteorological Administration, Beijing 100081, China
4 State Key Laboratory of Plateau Ecology and Agriculture, Qinghai University, Xining 810000, China
* Correspondence: sunam@cug.edu.cn

Abstract: Increasing of evidence suggests that rapid urbanization modifies precipitation and increases atmospheric instability in urban areas mainly due to urban heat island (UHI)-induced thermal effects and urban building-induced dynamic effects. However, few studies focus on the impact of urbanization on the diurnal variations in urban rainfall in terms of initiating time, frequency, amount, and intensity. Here, six years of hourly data from a relatively dense meteorological observation network in the Wuhan area in central China are used to quantify the issue of urbanization-induced diurnal variations in hourly rainfall and short-duration rainfall (SDR) events by comparing urban and surrounding rural stations during the summer dry period (13 July to 4 August), corresponding to hot weather with little rainfall and high evaporation. The results indicate that: (i) A higher frequency of hourly rainfall appears during night-time and afternoon over urban areas as compared to rural areas. The urban rainfall is more concentrated during the night-time; (ii) More and stronger urban SDR events are triggered during the night-time over urban areas, especially in terms of amount and intensity of rainfall events. A greater diurnal urban–rural difference in SDR events is detected than that in hourly mean rainfall; (iii) There is an enhancement in the night-time and afternoon urban SDR events when UHI intensity increases before they are initiated, and this is detectable in the frequency, amount, and intensity of SDR events. It is also found that the UHI-induced thermal effects are the main reason for the stronger nocturnal SDR events in Wuhan during the summer dry period, which is caused by increases in the convection current and water vapor flux convergence in the urban areas.

Keywords: urbanization; short-duration rainfall; diurnal variation; China; urban heat island intensity

1. Introduction

Early in the 1920s, Horton [1] observed some thunderstorms generated from cities in the northeast United States. Subsequently, a major North American field program, the Metropolitan Meteorological Experiment (METROMEX) [2,3], and high-density weather station networks and historical climate data series in a couple of major cities [4–7], provided more observational evidence that urban areas can significantly change precipitation due to human activities [4,8–10]. A recent review of 48 papers concerning the impact of urbanization on precipitation by Liu and Niyogi [11] indicated that the impacts of urbanization-induced rainfall modification are mostly seen in urban areas and places downwind of them, with an ~18% increase in the mean precipitation over the downwind areas and ~16% increase over the urban centers.

Several mechanisms have been proposed to explain urbanization-induced modifications using both observational and modeling approaches. These mechanisms include: (i) atmospheric destabilization and convection enhancement caused by Urban Heat Island

(UHI)-induced thermal perturbation over urban areas [5,12,13]; (ii) barrier effects caused by increased mechanical turbulence associated with increased surface roughness in the urban environment and thermal effects from urban canyons [14–16]; (iii) anthropogenic aerosols changing the microphysical processes in urban clouds [17–19], which may be crucial in urbanization-induced rainfall once the convection is triggered [19]. One or more mechanisms may dominate the urbanization-induced impacts over various cities with different geographical features [11].

The spatial characteristics of urbanization-induced rainfall are a research hotspot. Destabilization of the urban boundary layer and resulting downstream translation of the UHI circulation by synoptic flow [20] may be the reason for increased rainfall downwind of urban areas. A combination of low wind and high UHI can enhance convergence into the city leading to higher rainfall totals over urban areas [21]. Liu and Niyogi [11] indicated that modeling studies suggested an increase in rainfall over urban areas, while case studies showed a preference for increasing rainfall downwind of urban areas. Furthermore, several studies reveal that short-duration rainfall (SDR) is significantly more impacted by UHI than the normal rainfall during the summer in China [6,22,23]. For instance, Liang and Ding [22] found the occurrence frequency and intensity of hourly heavy precipitation show a phenomenon of "urban rain-island" in Shanghai, China. Wu et al. [23] also found that an urbanization-related positive trend is more related to SDR than to rainfall with a longer duration in the Pearl River Delta region of south China.

The diurnal variation of rainfall is essentially governed by the interaction of thermodynamic forcing, cloud microphysics, and local circulation [24,25]. Oki and Musiake [25] showed that evening convective rainfall is caused by thermodynamic forcing in inland areas, while morning low-level convergence is due to the interactions of local land-sea breezes and predominant monsoon winds in the coastal regions of Japan and Malaysia. Zhang and Zhai [26] revealed that surface solar radiative heating results in the maximum frequencies of extreme rainfall in the afternoon and/or early evening over most of China. Huff and Vogel [27] indicated that the evening maximum rainfall downwind of urban areas is associated with the interaction between UHI and atmospheric processes, which enhances the regional nocturnal thunderstorm anomalies. Recently, more studies have focused on the impacts of urbanization on the diurnal variation of rainfall using high spatial and temporal resolution rainfall data. Case studies showed that both mean and extreme rainfall have increased during the late afternoon and/or early night-time over and downwind of urban areas [28]. For example, the diurnal variation in rainfall intensity shows a significant urban-rural difference during the late afternoon and early night-time in the downtown areas of Shanghai, China, as a result of urbanization, local circulation associated with solar heating, and sea breezes [22].

The nocturnal Urban Heat Island Intensity (UHII) peak might be an important factor that causes the convergence of night-time propagating storms over inland cities such as Minneapolis, MN, Washington, DC [13], and Atlanta, GA [12]. It may also play a role in the increased frequency of storms over cities, such as Chicago, IL, New Orleans, LA, and Los Angeles, CA [4]. The combined influence of topography and urban thermal processes is the main cause of most mesoscale torrential rains that mostly occur in the evening or early morning in China [29]. Further, Ganeshan and Murtugudde [30] used a modeling approach over Minneapolis, MN, and determined that frictional convergence due to rough urban surfaces, aided by the nocturnal UHI, appears to enhance local convergence and attract propagating storms towards the urban center. On the other hand, the UHI produces a low-pressure anomaly causing local convergence during calm atmospheric conditionsand even the development of new convection [12,31].

After changing the surface parameters to typical urban surface values in a mesoscale model, Guo et al. [14] showed that convective cells were produced earlier over urban areas as a result of increased sensible heat flux. Additionally, Zhong and Yang [32] determined that the thermal effects of UHI could slow cold front movement and thereby delay the occurrence of peak rainfall over urban areas. However, all studies focused on the urbanization-induced diurnal variation of rainfall variations based on hourly rain rates. The initiating time of rainfall is crucial in urban rainfall forecasts, especially for extreme rainfall events since such extreme weather directly increases risks in social, economic, and human systems. Exploring the effects of urbanization on rainfall initiating time can help improve rainfall forecasts, and it is also important for investigating more detailed characteristics of urbanization-induced rainfall events.

The summer dry period, corresponding to hot weather with little rainfall and high evaporation [33,34], usually occurs from mid-July to mid-August. It is one of the most relevant disasters characterized by rising temperature and more extreme precipitation [35]. It not only affects agricultural production and food security, but also threatens the economy and environment. In addition, rainfall frequency during the summer dry period is mainly dominated by SDR in the Wuhan area. The SDR events during this period have great implications for regional agricultural production, water resources, and urban operation and management. Thus, the causes and prediction of the summer dry period have received much attention. However, less attention has been paid to the influence of urbanization-induced diurnal variations in SDR events during the summer dry period.

By comparing the difference between urban and rural rainfall variables during the summer dry period in Wuhan, especially focusing on the SDR events, the possible influences of urbanization on diurnal variations in rainfall and SDR are explored. The remainder of this paper is arranged into five parts. The materials and methodologies used are introduced in Section 2. Section 3 presents the diurnal difference between urban and rural rainfall and SDR, and the possible impacts of UHI-induced thermal effects. Concluding remarks and discussions are presented in Section 4.

2. Materials and Methodologies

2.1. Data

The dataset of hourly rainfall and SAT over the Wuhan region was obtained from the National Meteorological Information Center (NMIC) of the China Meteorological Administration (CMA). These data have been extensively used to investigate the characteristics of sub-daily precipitation in China [23,36,37]. A total of 123 stations with hourly rain gauge records and hourly SAT data are established in the Wuhan area, and such relatively high density is beneficial for a more accurate comparison between urban and rural stations. The double tipping-bucket telemetering rain gauge is widely used in automatic weather stations in China. The rain gauges carry out hour-by-hour rainfall observations in the study area [38–40]. The study mainly focuses on the summer dry period from 2012 to 2018. However, 2015 was excluded due to missing values exceeding 50%.

Daily precipitation for the middle and lower regions of the Yangtze River, hourly wind vector field, specific humidity, and geopotential height for Jianghan Plain form ERA5 reanalysis data, provided by the European Centre for Medium-Range Weather Forecasts (ECMWF), are used in this study. The spatial resolution for these data is $0.25^\circ \times 0.25^\circ$. Daily precipitation data are available from 1980 to 2010. The 925 hPa wind is presented as the prevailing wind, and the prevailing wind in Wuhan during the summer dry period is from the southwest (Figure 1c). The 925 hPa divergence, wind vector, and specific humidity were used to calculate the water vapor flux divergence at low level. The area with a geometric height of less than 120 m is classified as Jianghan Plain, and it is not shown in the Figures.

The night-time light images were obtained from the Defense Meteorological Satellite Program/Operational Linescan System (DMSP/OLS) for the period 2000–2013, which needs to be adjusted to set thresholds in the city map in different cities with inconsistent urban development processes. It was pointed out that the area with the highest consistency was found in the Wuhan region when the selection threshold was set at 53% [41]. Thus, in this paper, the night-time light area greater than or equal to 53% over the Wuhan area is taken to be the Wuhan urban areas. At this value, the interference of transient light sources can be eliminated, and the integrity of the central part can be ensured. In addition, the land cover/land use (LULC) dataset was obtained from the Resources and Environmental

Sciences Data Center, Chinese Academy of Sciences, at 1 km resolution. Six land-use types were classified in the LULC data, including cultivated land; forestry land; grassland; water area; urban, rural and residential land; and unused land.

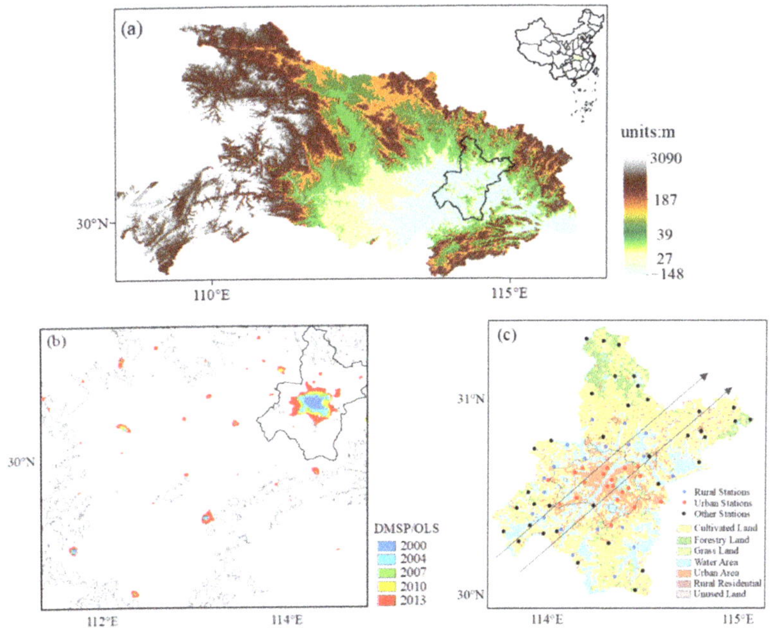

Figure 1. Topography of Hubei Province (**a**) and night-time light of Jianghan Plain (**b**). The solid black line delineates the Wuhan area. Land use and distribution of stations in the Wuhan region (**c**). The unit of the terrain is m, and the black line indicates the location of Wuhan in (**a**); the solid grey line in (**b**) is the 100 m elevation contour line, and the thin black solid line in (**c**) delineates the Wuhan urban areas based on night-time lighting. The dots in (**c**) represent 76 stations in Wuhan, with red dots for urban stations, blue dots for rural stations, and black dots for the rest.

2.2. Study Area

Studies have shown that different cities with different backgrounds (including climate regimes, terrain, and urbanization level, etc.) and urban expansion may have different characteristics and dominant mechanisms [13,42–45]. As the capital of Hubei Province, Wuhan has experienced the strongest ongoing and accelerated urbanization during the past few decades in the Jianghan Plain, which is located in the center of China (Figure 1a). The rapid urbanization can be seen from the changes in several indicators, such as night-time light intensity (Figure 1b) and urban land use data (Figure 1c). The population density shows an increase of 61% in 2015 compared to 1990, and the urban land use ratio has tripled from 1980 to 2018. Additionally, a significant UHI has been found over the past few decades using both annual and seasonal mean Surface Air Temperature (SAT) time series at the Wuhan station from the basic national meteorological observation network of China [46,47]. Furthermore, the land surface temperature in urban areas has increased [48,49]. Wuhan is not only one of many rapidly growing cities in central China but also a typical subtropical inland megacity. Accompanied by this rapid urbanization, a relatively dense meteorological observation network in and around the Wuhan areas was established to provide hourly and synoptic observations for regular weather parameters. Hence, Wuhan could serve as a great case-study site to examine the potential urbanization-induced impacts on rainfall characteristics on the diurnal time scale.

2.3. Methods

2.3.1. Definition of the Summer Dry Period

To define the summer dry period, the time series of daily precipitation is obtained by smoothing five points from 1 June to 15 September (Figure 2), using the reanalysis data of precipitation in the middle and lower reaches of the Yangtze River (27.5–32.5° N, 112–123° E) during 1981–2010. The daily precipitation was analyzed with standardized anomalies. The summer dry period begins with negative daily precipitation for 7 consecutive days and ends with daily precipitation greater than zero for 1 day. Therefore, in this paper the summer dry period is defined to run from 13 July to 4 August (Figure 2).

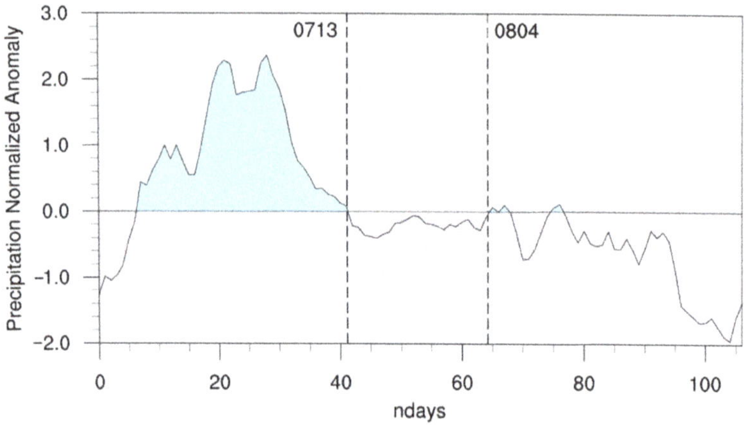

Figure 2. Standardized anomalies of daily mean precipitation in the middle and lower reaches of the Yangtze River basin from 1 June to 15 September. The blue/pink areas are above/below 0.

2.3.2. Classification of Urban and Rural Stations

Although many quality tests have been applied by NIMC, further quality control regarding missing values still needs to be performed. Therefore, 76 observational stations are retained in the study area after excluding stations with more than 5% of hours of missing data per year (Figure 1c).

As has been pointed out, the urban land use areas within the 4-km radius buffer zone around the station have the highest correlation with the SAT trend [50]. The areal percentage of urban land use in the buffering circle with a 4 km radius around the station was used as an urbanization indicator. All stations were ranked in order of urban land-use ratio, and the top 20 were selected as urban stations, among which the lowest urbanization percentage was 25%. Therefore, an urban land-use ratio of 25% within the buffer zone is taken as the standard to distinguish urban and rural stations.

In addition, the elevation range of urban stations is 12–44 m, with an average elevation of 29.75 m. Considering the fact that the comparison of urban and rural stations will be influenced by wind direction, elevation, and distance from the urban center, the conditions for the selection of rural station samples were defined as follows: (i) the elevation fluctuation range does not exceed the average elevation of urban stations by 20 m; (ii) the distance from the city boundary is less than or equal to 20 km; (iii) stations should be evenly distributed around the city. Finally, 20 urban and 20 rural stations were selected in this paper, and the distribution is shown in Figure 1c.

2.3.3. Definition of Rainfall Variables

Different definitions and procedures of rainfall event separation can be found in previous studies [51,52]. With specific constraints of rainfall characteristics, SDR can be defined based on the obtained rainfall events. In this study, the SDR begins with hourly rainfall exceeding 0.1 mm and ends with hourly rainfall falling to 0.1 mm, with a duration equal to or less than 6 h, following Yang [6]. The rainfall timing is defined as the rainfall initiating time in terms of Local Time (LT). For each rainfall event, the rainfall amount was defined as the sum of hourly rainfall, the rainfall duration was defined as the period time from start to end, and the rainfall intensity was calculated as the ratio of rainfall amount to rainfall duration. Frequency and average amount of hourly rainfall, frequency, average amount, and intensity of SDR events, respectively, were also analyzed. The diurnal variation in these variables of hourly rainfall and SDR were calculated using standardized approaches.

2.3.4. Definition of the Urbanization Indicator

In this paper, the urban SDR events are classified under different UHI backgrounds before they are initiated, i.e., SDR with strong UHI and SDR with weak UHI, respectively.

The urbanization indicator Urban Heat Island Intensity Anomaly (*UHIIA*) is defined by Wu et al. [23]:

$$UHII = T_U - T_R \tag{1}$$

$$\overline{UHII} = \sum_{2012}^{2018} UHII/6 \tag{2}$$

$$UHIIA = UHII - \overline{UHII} \tag{3}$$

For these equations, *UHII* represents the difference between hourly SAT over urban areas (T_U) and hourly SAT over rural areas (T_R), and $\overline{UHII}$ is calculated by averaging *UHII* at the same LT in 2012–2018. Therefore, *UHIIA* values greater/less than 0 are considered positive/negative anomalies of *UHII*.

Five hourly *UHIIA* values prior to the timing of all rainfall events are first selected, and then the events with a positive mean *UHIIA* within five hours prior to the rainfall are defined as the events with strong UHI; the remaining events, that is, the events with a negative mean *UHIIA* within five hours prior to the rainfall timing, are classified as events with weak UHI.

To clarify, rainfall or SDR initiated during 0000–0600 LT, 0600–1200 LT, 1200–1800 LT and 1800–0000 LT is referred to as the night-time, morning, afternoon and early night rainfall, respectively.

3. Results

3.1. Diurnal Urban-Rural Differences in Rainfall Variables

The diurnal variations in the frequency and average amount of hourly rainfall at the urban and rural stations, respectively, are examined during the summer dry period over the Wuhan area (Figure 3a–c). The diurnal variation in frequency exhibits a bimodal structure, with the first peak during 0500–0800 LT, and the second one during 1300–1500 LT, but it is more concentrated in the first peak. However, a trimodal structure is seen in the diurnal variation in amount, with the peaks at 0400–0800 LT, 2000–2100 LT, and 1300–1400 LT, respectively, and the amount is also more concentrated in the first two peaks.

Furthermore, Figure 3c,d shows the diurnal urban-rural differences in the frequency and average amount of hourly rainfall. During 2200–0600 LT and 1400–1900 LT, a higher frequency appears over Wuhan urban areas than rural areas; however, there is no obvious diurnal urban-rural difference in the average amount.

Spatially, more hourly rainfall appears over and downwind of Wuhan urban areas, but it is more concentrated over urban areas during the night-time (Figure 4a,b). Similarly, there is also no obvious spatial difference in amount between urban and rural areas (Figure 4c,d).

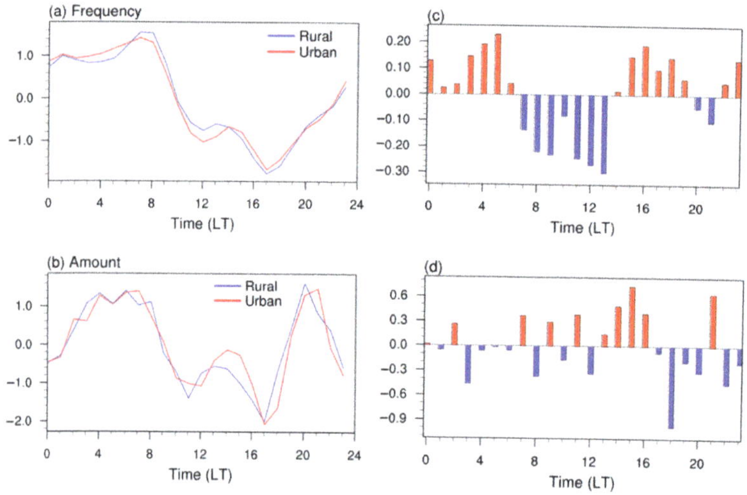

Figure 3. The diurnal variations in rainfall frequency (**a**,**c**), average amount (**b**,**d**) at the urban and rural stations (**a**,**b**), and their difference between urban and rural stations (**c**,**d**). Red/blue bar indicates the positive/negative difference, respectively.

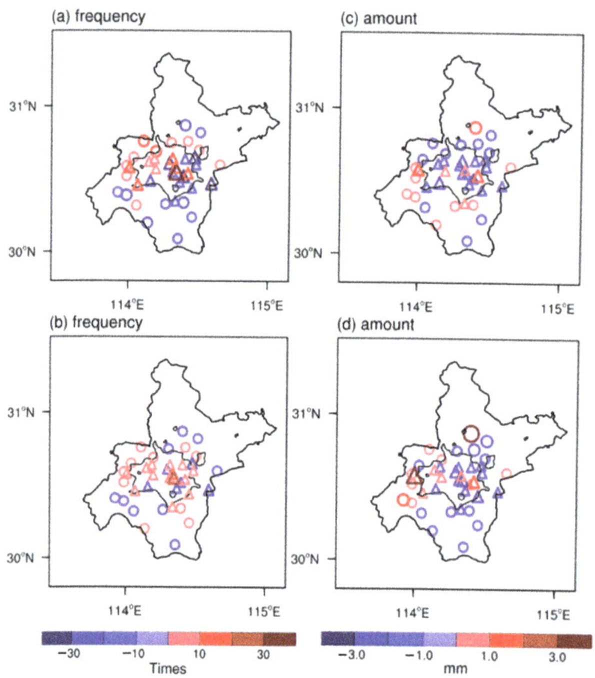

Figure 4. Spatial distribution of rainfall frequency (**a**,**b**), average amount (**c**,**d**) anomaly in Wuhan area (**a**,**c**) and night-time (0000–0600 LT) (**b**,**d**), respectively. The units of rainfall frequency and amount are times and mm; the urban stations are shown as triangles, and the rural stations are shown as hollow circles; the thick black solid line delineates the Wuhan area, and the thin black solid line delineates the urban areas.

3.2. Diurnal Urban-Rural Differences in SDR Variables

A total of 743 and 736 SDR events were collected at the 20 urban stations and 20 rural stations, respectively, over the Wuhan area during the study period. The average amounts and intensity of urban SDR events are 5.32 mm and 2.35 mm/h, and those of rural SDR events are 5.80 mm and 2.62 mm/h, respectively.

The diurnal variations in frequency, amount, and intensity of SDR with different initiating times at urban and rural stations are shown in Figure 5a–c. The diurnal variations in the three variables of SDR events all reveal a bimodal structure, with the first peak during 0500–0800 LT, the second peak during 1300–1500 LT, and the valley at 1700 or 1800 LT. However, the diurnal variation in frequency is more concentrated on the first peak when compared to the amount and intensity (Figure 5a–c).

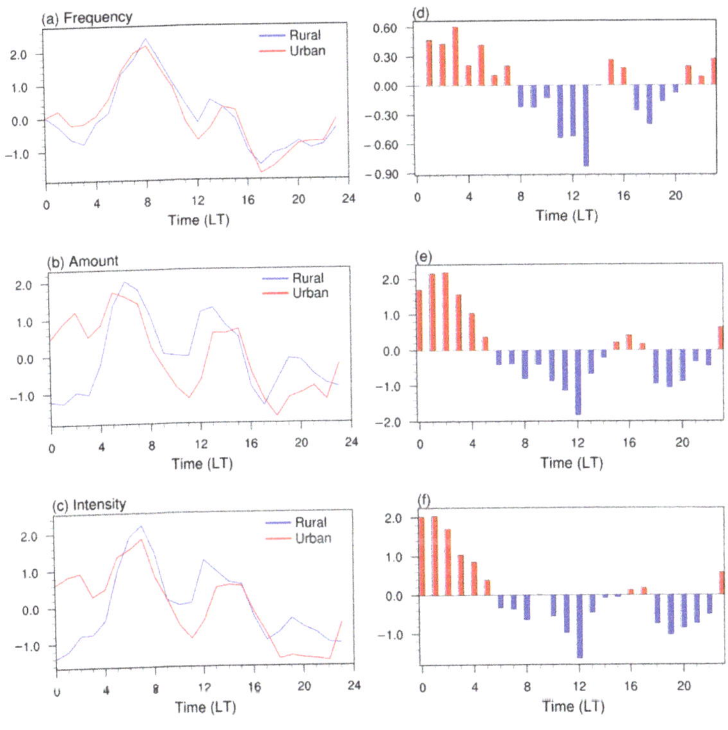

Figure 5. The diurnal variations in SDR frequency (**a**,**d**), average amount (**b**,**e**), and average intensity (**c**,**f**) at urban and rural stations (**a**–**c**), and their differences between urban and rural stations (**d**–**f**). Red/blue bar indicates the positive/negative difference, respectively.

To further understand the urban-rural difference of SDR variables at different initiating times, their diurnal difference changes were calculated (Figure 5d–f). More and stronger SDR events are triggered during the night-time over Wuhan urban areas, corresponding to higher rainfall frequency, amount, and intensity. Except for the night-time, the urban SDR events initiated from 1400 to 1600 LT also show greater variability. When compared to the diurnal urban-rural difference in the hourly rainfall variables, a greater urban-rural difference is seen in the SDR variables, with a greater amount and intensity during the night-time (Figures 3a–d and 5d–f). Thus, the urban environment may lead to a greater occurrence of the night-time SDR in the Wuhan area during the summer dry period.

The spatial distributions of frequency, amount, and intensity of SDR events are shown in Figure 6, respectively. An obviously greater amount and intensity are exhibited over and downwind of the urban areas, but it is also more concentrated in urban areas during the night-time (Figure 6c–f). However, there is no significant spatial urban-rural difference in SDR frequency (Figure 6a,b). Therefore, a greater urban-rural difference appears in the amount and intensity of SDR events over the Wuhan area during the study period, especially compared with hourly rainfall.

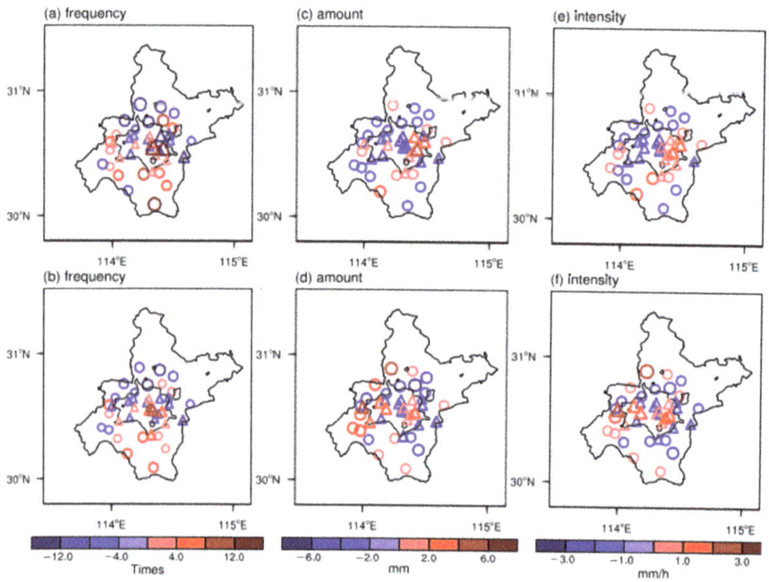

Figure 6. Spatial distribution of SDR frequency (**a**,**b**), average amount (**c**,**d**), and average intensity (**e**,**f**) anomaly in Wuhan area (**a**,**c**,**e**) and night-time (0000–0600 LT) (**b**,**d**,**f**). The units of rainfall frequency, amount, and intensity are times, mm, and mm/h, respectively; the urban stations are shown as triangles, and the rural stations are shown as circles; the thick black solid line delineates the Wuhan area, and the thin black solid line delineates the urban areas.

3.3. UHI-Induced Impacts on SDR

The peak of diurnal variation appears at the night-time over the Wuhan urban area (Figure 7). Therefore, the impacts of nocturnal UHI-induced thermal perturbation on favorable conditions for rainfalls are examined.

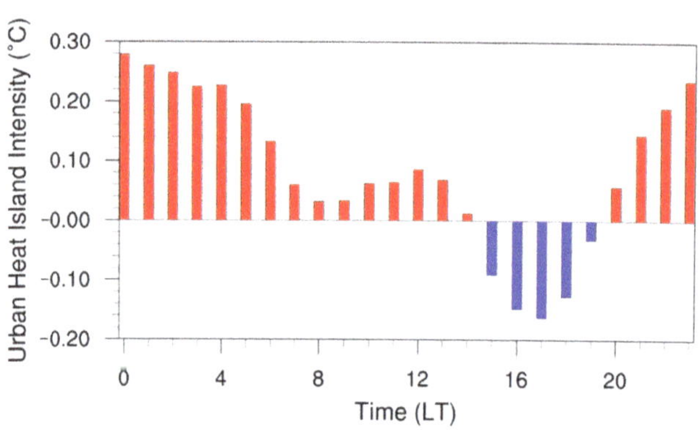

Figure 7. Diurnal variation of the urban heat island intensity (UHII) anomaly in the Wuhan area. The unit is °C. Red/blue bar indicates the positive/negative UHII, respectively.

A total of 572 urban SDR events are collected at the urban stations during the study period. They are classified further into 262 SDR events with strong UHI and 310 SDR with weak UHI, based on the *UHIIA* within the six hours prior to the initiating rainfall time. The average amount and intensity of SDR with strong UHI are 3.67 mm and 1.62 mm/h, and those with weak UHI are 3.01 mm and 1.43 mm/h, respectively.

A bimodal structure is exhibited in the diurnal variations of SDR frequency with different UHI backgrounds, in which the peaks of SDR with strong UHI appear at 0800 LT and 1400 LT, while those with weak UHI appear at 0900 LT and 1500 LT, respectively (Figure 8a). Nevertheless, a completely different diurnal change is exhibited under different UHII background in the amount and intensity of SDR (Figure 8b,c). The amount and intensity of SDR with strong UHI also exhibit two peaks, which are 0300 LT and 1300 LT. However, a unimodal structure appears in the diurnal variation of the amount and intensity of SDR with weak UHI, with the maximum at 0700 LT. In addition, the diurnal variation of urban SDR events with strong UHI exhibit an earlier shift during the night-time than that of events with weak UHI (Figure 8a–c).

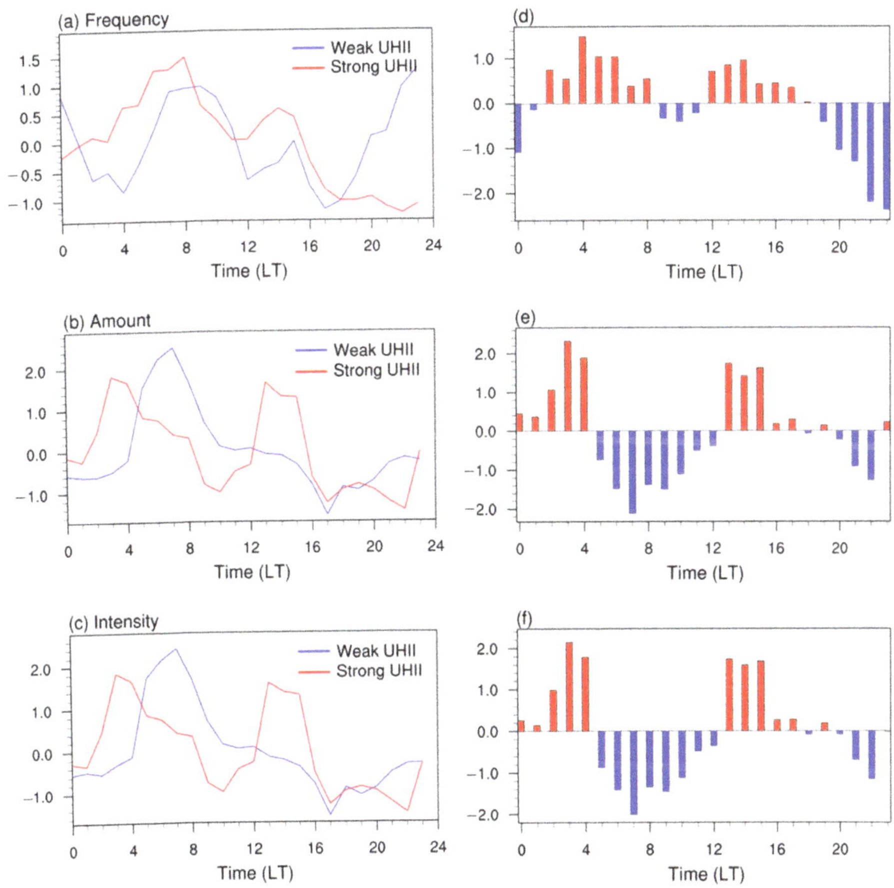

Figure 8. Diurnal variations of frequency (**a**), amount, (**b**) and intensity (**c**) of SDR with strong UHI, and weak UHI, and their difference between SDR with strong UHI and weak UHI (**d**–**f**). Red/blue bar indicates the positive/negative difference, respectively.

Figure 8d–f shows the different diurnal variations between SDR with strong UHI and weak UHI. The SDR with strong UHI shows a greater amount and intensity during 0000–0400 LT and 1300–1700 LT, especially compared to frequency. Therefore, an obvious greater enhancement in the SDR amount and intensity during the night-time and afternoon is detected in Wuhan urban areas, with an increasing UHI before the SDR initiated.

The spatial distribution of the difference in variables between SDR with strong UHI and weak UHI shows that nearly 90% of urban stations show a positive amount and intensity, and 70% of them passed the test of statistical significance at the 90% confidence level (Figure 9c,e). Under the southwest winds in the lower atmosphere, a significant enhancement is observed in downwind portions of the urban areas, with an increasing UHI before the SDR initiated. Especially during the night-time, 95% and 100% of urban stations show positive SDR amount and intensity differences between strong UHI and weak UHI events, respectively (Figure 9d,f). This positive difference is obviously greater at night-time when compared to all events, and it is distributed throughout the urban areas.

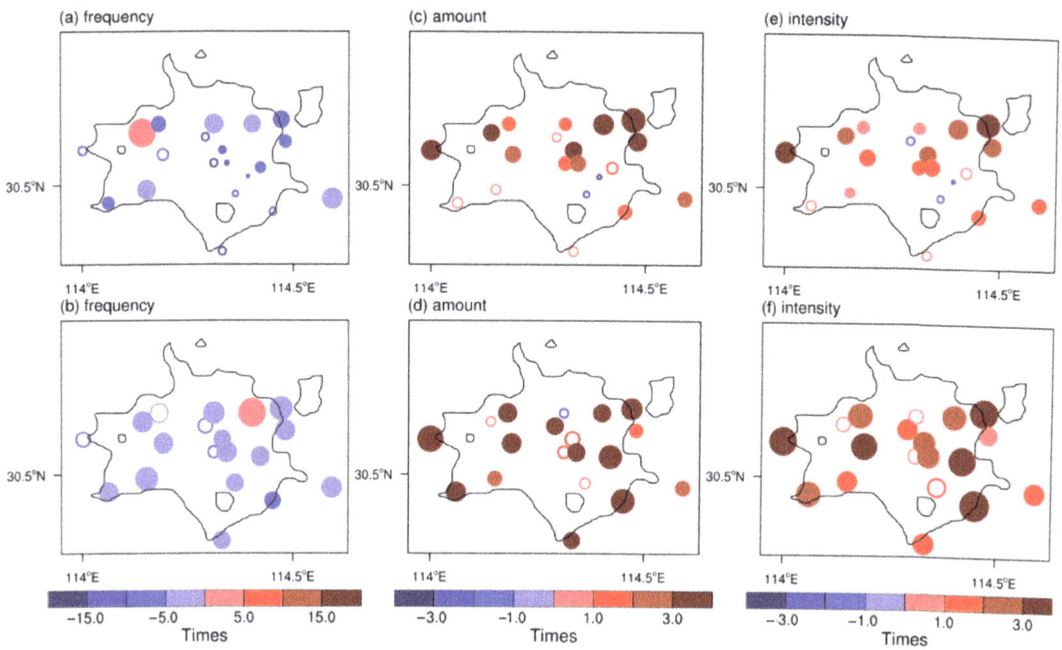

Figure 9. Spatial distribution of different frequency (**a**), average amount (**b**) and average intensity (**c**) between SDR with strong UHI, and weak UHI in the Wuhan area (**a**,**c**,**e**) and night-time (0000–0600 LT) (**b**,**d**,**f**), respectively. The units of rainfall frequency, amount, and intensity are times, mm, and mm/h, respectively; the solid circles/hollow circles indicate statistically significant/insignificant at 90% confidence levels; the black solid line delineates the urban areas.

Furthermore, the spatial distributions of the mean SAT within 5 h prior to the initiating time of urban SDR events with strong UHI, weak UHI, and their difference are shown in Figure 10. A greater warm center of the SAT difference is observed over and downwind of the urban areas between the SDR events with strong UHI and weak UHI. Moreover, the warmer area generally corresponded to stronger SDR, when comparing the amount and intensity of urban SDR events with strong UHI and weak UHI (Figures 9c,e and 10c).

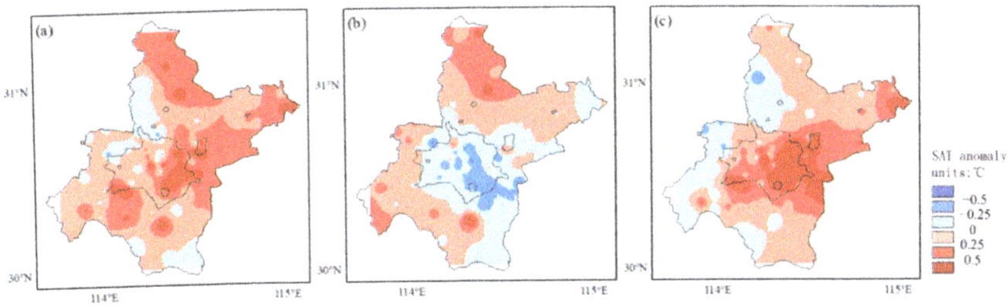

Figure 10. Spatial distribution of average SAT anomaly in five hours before the SDR with strong UHI (**a**), SDR with weak UHI (**b**), and their difference between SDR with strong UHI and weak UHI (**c**). The unit is °C.

The water vapor flux divergence at 925 hPa at 0200 LT in the Jianghan Plain shows that a nocturnal convergence appears over the center of the Jianghan Plain (Figure 11a). The nocturnal water vapor convergence might explain the aforementioned greater hourly rainfall and SDR at night-time over Wuhan (Figures 3 and 5). However, the nocturnal convergence appears over and downwind of Wuhan urban areas when the urban SDR events are affected by strong UHI during the night-time (Figure 11b). This indicates that the UHI-thermal effect concentrates the widespread nocturnal water vapor flux convergence zone on the Wuhan urban areas, and enhances the urban-rural difference, which is conducive to developing and maintaining urban rainfall events during night-time.

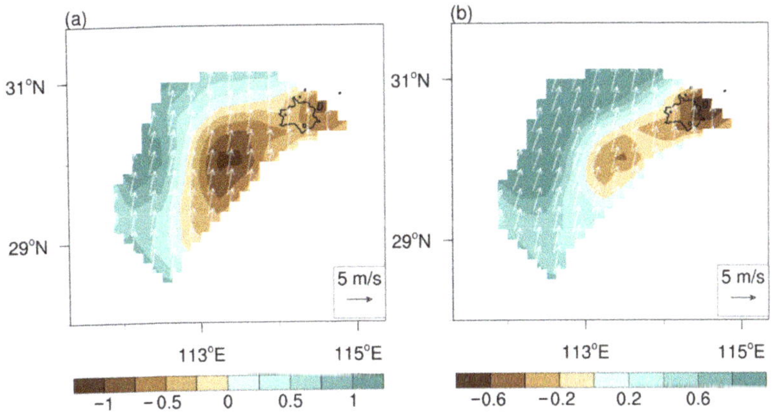

Figure 11. Spatial distribution of water vapor flux divergence and wind at 925 hPa in the Jianghan Plain at 0200 LT when appears night-time urban SDR events (**a**), and when affected by a strong urban heat island before the night-time urban SDR events initiated (**b**). The unit of water vapor flux divergence and wind vector are g $\times$cm^{-2}hPa^{-1}s^{-1} and m s^{-1}, respectively.

4. Discussion

The diurnal variations of SDR events also show a bimodal structure. The morning peak may be caused by the interaction between the nocturnal boundary layer low-level jet, aligned with the southwesterly synoptic-scale flow, which carries abundant moisture, and the mechanical lifting by the windward side of the Dabie mountain area, which is located downwind Wuhan [53]. The afternoon peak may be due to the surface heating [54].

There are some uncertainties in this work. First, only a specific period of a few years is considered in this study, although many rainfall variables of summer rainfall events during the entire study period have been analyzed. It is well known that rainfall events vary with different large-scale circulation and/or mesoscale systems. Other processes may also play an important role in explaining the diurnal urban-rural differences in rainfall events besides the UHI-induced thermal perturbation effect highlighted in this work. For example, Ganeshan and Murtugudde [30] suggested that the urban frictional convergence, when accompanied by a nocturnal UHI, can significantly modify the spatial position of a storm during night-time. The combined influence of topography and the urban thermal process can increase summer rainfall frequencies in the evening or early morning [29]. It was mentioned by Fu [55] that the intensified rainfall zone of each city cluster is different because of their unique geography and structure. Li [56] also found that terrain effects play an important role in affecting rainfall frequency, amount, and intensity. Therefore, UHI-induced thermal perturbation may not be the only and/or direct cause of the diurnal urban-rural differences in rainfall events. The complex interactions between land surface, planetary boundary layer, aerosols, atmospheric dynamics, and thermodynamic processes under different environmental and atmospheric conditions may help to explain the detailed differences in rainfall modification in various urban areas.

Second, the diurnal variations in rainfall events over neighboring non-urban areas are taken as the reference group to quantify the urbanization impacts on urban areas for this study. This urban-rural difference method assumes that both areas share the same meteorological conditions. However, Wuhan is surrounded by mountains with complex terrains, which will introduce some differences between urban and rural areas even without the impacts of urbanization. To minimize these uncertainties and further our study, in future work mesoscale models may be run to quantify the urbanization impacts on rainfall events and explore detailed the physical processes involved.

Although there are a few questions which need to be addressed in the future, the key conclusions drawn in the study are robust, and these are of importance to the planning and management of the city. In particular, the bimodal structure characteristic of hourly rainfall and the enhancement of the urban SDR at night-time and afternoon due to urbanization effects could be considered in designing and managing the urban sewage infrastructure so as to drain the urban floods in a more timely manner during the peaks of hourly precipitation frequency and intensity.

5. Conclusions

This paper aims to understand the urbanization-induced variations in diurnal short-duration rainfall events in Wuhan, an inland city in China, by using the hourly surface observed precipitation data during the summer dry period in 2012–2018. The main findings can be summarized as follows:

1. The diurnal variation in the frequency of hourly rainfall shows a bimodal structure over the Wuhan area, with the peaks at night-time (2200–0600 Local Time) and afternoon (1400–1900 Local Time), respectively. More urban rainfall is concentrated at night-time and afternoon, which differs from the pattern of rural rainfall. Obviously greater frequency is exhibited over and downwind of Wuhan urban areas but is more concentrated on the urban areas during the night-time;
2. The diurnal variations of SDR events also reveal a bimodal structure over the Wuhan area, with peaks in the early morning (0600–1200 Local Time) and afternoon. An enhancement in urban SDR events can be seen during the night-time when compared to the rural events, especially in amount and intensity. More numerous and intense SDR events occur over and downwind of urban areas but they are also more concentrated on the urban areas during the night-time. A greater diurnal urban-rural difference in SDR variables is detected than that in hourly rainfall during the dry summer over the Wuhan area;
3. An obvious difference exists in the diurnal variations in urban SDR events when affected by different urban heat islands before the event initiated. There is an enhancement in the night-time and afternoon SDR events when urban heat islands increased, and this is detectable in the frequency, amount, and intensity of SDR events. Over 95% of urban stations show a significantly positive difference in SDR amount and intensity

between strong UHI and weak UHI events, and they are distributed throughout the urban area. The warmer area before urban SDR initiated corresponded to the area of stronger SDR when comparing urban SDR events with strong UHI and weak UHI; The nocturnal UHII peak enhances the positive urban-rural difference in water vapor flux convergence, which is essential to the triggering, development, and maintenance of SDR.

Author Contributions: G.R. contributed the central idea, G.R., S.K.T. and Y.M. designed the study, Y.M. performed the calculations and wrote the main manuscript, G.R. and S.K.T. revised the manuscript. All authors have read and agreed to the published version of the manuscript.

Funding: This research was funded by the National Key R&D Program of China (2018YFA0605603).

Data Availability Statement: The data used in the manuscript are publicly available for ERA-5 (https://www.ecmwf.int/en/forecasts/dataset/ecmwf-reanalysis-v5 (accessed on 10 March 2021)), DMSP/OLS (https://ngdc.noaa.gov/eog/dmsp/downloadV4composites.html (accessed on 10 March 2021)), and the LULC from the Resources and Environmental Sciences Data Center, Chinese Academy of Sciences (https://www.resdc.cn/data.aspx?DATAID=264 (accessed on 10 March 2021)).

Conflicts of Interest: The authors declare no conflict of interest.

References

1. Horton, R.E. Thunderstorm-Breeding Spots. *Mon. Weather Rev.* **1921**, *49*, 193. [CrossRef]
2. Changnon, S.A.; Huff, F.A.; Semonin, R.G. METROMEX: An Investigation of Inadvertent Weather Modification. *Bull. Am. Meteorol. Soc.* **1971**, *52*, 958–968. [CrossRef]
3. Shepherd, J.M. A Review of Current Investigations of Urban-Induced Rainfall and Recommendations for the Future. *Earth Interact.* **2005**, *9*, 1–27. [CrossRef]
4. Huff, F.A.; Changnon, S.A. Precipitation Modification by Major Urban Areas. *Bull. Am. Meteorol. Soc.* **1973**, *54*, 1220–1232. [CrossRef]
5. Shepherd, J.M.; Pierce, H.; Negri, A.J. Rainfall Modification by Major Urban Areas: Observations from Spaceborne Rain Radar on the TRMM Satellite. *J. Appl. Meteorol.* **2002**, *41*, 689–701. [CrossRef]
6. Yang, P.; Ren, G.; Yan, P. Evidence for a Strong Association of Short-Duration Intense Rainfall with Urbanization in the Beijing Urban Area. *J. Clim.* **2017**, *30*, 5851–5870. [CrossRef]
7. Yonetani, T. Increase in Number of Days with Heavy Precipitation in Tokyo Urban Area. *J. Appl. Meteorol.* **1982**, *21*, 1466–1471. [CrossRef]
8. Huff, F.A. Urban effects on the distribution of heavy convective rainfall. *Water Resour. Res.* **1975**, *11*, 889–896. [CrossRef]
9. Landsberg, H.E. Man-Made Climatic Changes: Man's activities have altered the climate of urbanized areas and may affect global climate in the future. *Science* **1970**, *170*, 1265–1274. [CrossRef]
10. Huff, F.A.; Changnon, S.A. Climatological Assessment of Urban Effects on Precipitation at St. Louis. *J. Appl. Meteorol.* **1972**, *11*, 823–842. [CrossRef]
11. Liu, J.; Niyogi, D. Meta-analysis of urbanization impact on rainfall modification. *Sci. Rep.* **2019**, *9*, 7301. [CrossRef]
12. Bornstein, R.; Lin, Q. Urban heat islands and summertime convective thunderstorms in Atlanta: Three case studies. *Atmos. Environ.* **2000**, *34*, 507–516. [CrossRef]
13. Ganeshan, M.; Murtugudde, R.; Imhoff, M.L. A multi-city analysis of the UHI-influence on warm season rainfall. *Urban Clim.* **2013**, *6*, 1–23. [CrossRef]
14. Guo, X.; Fu, D.; Wang, J. Mesoscale convective precipitation system modified by urbanization in Beijing City. *Atmos. Res.* **2006**, *82*, 112–126. [CrossRef]
15. Shem, W.; Shepherd, M. On the impact of urbanization on summertime thunderstorms in Atlanta: Two numerical model case studies. *Atmos. Res.* **2009**, *92*, 172–189. [CrossRef]
16. Niyogi, D.; Holt, T.; Zhong, S.; Pyle, P.C.; Basara, J. Urban and land surface effects on the 30 July 2003 mesoscale convective system event observed in the southern Great Plains. *J. Geophys. Res.* **2006**, *111*, D19. [CrossRef]
17. Rosenfeld, D. Suppression of rain and snow by urban and industrial air pollution. *Science* **2000**, *287*, 1793–1796. [CrossRef]
18. Rosenfeld, D.; Lohmann, U.; Raga, G.B.; O'Dowd, C.D.; Kulmala, M.; Fuzzi, S.; Reissell, A.; Andreae, M.O. Flood or drought: How do aerosols affect precipitation? *Science* **2008**, *321*, 1309–1313. [CrossRef]
19. Cotton, W.R.; van den Heever, S.C. Urban Aerosol Impacts on Downwind Convective Storms. *J. Appl. Meteorol. Climatol.* **2007**, *46*, 828–850. [CrossRef]
20. Baik, J.-J.; Kim, Y.-H.; Chun, H.-Y. Dry and Moist Convection Forced by an Urban Heat Island. *J. Appl. Meteorol.* **2001**, *40*, 1462–1475. [CrossRef]

21. Dou, J.; Wang, Y.; Bornstein, R.; Miao, S. Observed Spatial Characteristics of Beijing Urban Climate Impacts on Summer Thunderstorms. *J. Appl. Meteorol. Climatol.* **2015**, *54*, 94–105. [CrossRef]
22. Liang, P.; Ding, Y. The long-term variation of extreme heavy precipitation and its link to urbanization effects in Shanghai during 1916–2014. *Adv. Atmos. Sci.* **2017**, *34*, 321–334. [CrossRef]
23. Wu, M.; Luo, Y.; Chen, F.; Wong, W.K. Observed Link of Extreme Hourly Precipitation Changes to Urbanization over Coastal South China. *J. Appl. Meteorol. Climatol.* **2019**, *58*, 1799–1819. [CrossRef]
24. Ohsawa, T.; Ueda, H.; Hayashi, T.; Watanabe, A.; Matsumoto, J. Diurnal Variations of Convective Activity and Rainfall in Tropical Asia. *J. Meteorol. Soc. Japan. Ser. II* **2001**, *79*, 333–352. [CrossRef]
25. Oki, T.; Musiake, K. Seasonal Change of the Diurnal Cycle of Precipitation over Japan and Malaysia. *J. Appl. Meteorol.* **1994**, *33*, 1145–1163. [CrossRef]
26. Zhang, H.; Zhai, P. Temporal and spatial characteristics of extreme hourly precipitation over eastern China in the warm season. *Adv. Atmos. Sci.* **2011**, *28*, 1177–1183. [CrossRef]
27. Huff, F.A.; Vogel, J.L. Urban, Topographic and Diurnal Effects on Rainfall in the St. Louis Region. *J. Appl. Meteorol.* **1978**, *17*, 565–577. [CrossRef]
28. IPCC. Working Group III Contribution to the Fifth Assessment Report of the Intergovernmental Panel on Climate Change. In *Climate Change 2014: Mitigation of Climate Change*; Ottmar, E., Ramón, P.-M., Youba, S., Jan, C.M., Ellie, F., Susanne, K., Kristin, S., Anna, A., Ina, B., Steffen, B., et al., Eds.; Cambridge University Press: Cambridge, UK; New York, NY, USA, 2014.
29. Sun, J.; Yang, B. Meso-β Scale Torrential Rain Affected by Topography and the Urban Circulation. *Chin. J. Atmos. Sci.* **2008**, *6*, 1352–1364. (In Chinese)
30. Ganeshan, M.; Murtugudde, R. Nocturnal propagating thunderstorms may favor urban "hot-spots": A model-based study over Minneapolis. *Urban Clim.* **2015**, *14*, 606–621. [CrossRef]
31. Dixon, P.G.; Mote, T.L. Patterns and Causes of Atlanta's Urban Heat Island–Initiated Precipitation. *J. Appl. Meteorol.* **2003**, *42*, 1273–1284. [CrossRef]
32. Zhong, S.; Yang, X.-Q. Ensemble simulations of the urban effect on a summer rainfall event in the Great Beijing Metropolitan Area. *Atmos. Res.* **2015**, *153*, 318–334. [CrossRef]
33. Fang, J.; Bai, A.; Xiao, K.; Tian, W. Analysis of climatic characteristics and causes of the drought in Shaanxi. *Chin. J. Agric. Res. Arid Areas* **2009**, *27*, 28–34. (In Chinese)
34. Luo, B.; Zhang, C. Characteristics of droughts and floods in Hunan during the drought period and their correlation with the early circulation. *Chin. J. Meteorol. Sci. Technol.* **2009**, *37*, 19–24. (In Chinese)
35. Spinoni, J.; Naumann, G.; Carrao, H.; Barbosa, P.; Vogt, J. World drought frequency, duration, and severity for 1951-2010. *Int. J. Climatol.* **2014**, *34*, 2792–2804. [CrossRef]
36. Guo, J.; Su, T.; Li, Z.; Miao, Y.; Li, J.; Liu, H.; Xu, H.; Cribb, M.; Zhai, P. Declining frequency of summertime local-scale precipitation over eastern China from 1970 to 2010 and its potential link to aerosols. *Geophys. Res. Lett.* **2017**, *44*, 5700–5708. [CrossRef]
37. Li, J.; Yu, R.; Zhou, T. Seasonal Variation of the Diurnal Cycle of Rainfall in Southern Contiguous China. *J. Clim.* **2008**, *21*, 6036–6043. [CrossRef]
38. Li, Y.; Ren, Z.; Chen, G.; Miao, J.; Liu, F. Difference and correlation analysis between self-registered and automatically observed rainfall. *Chin. J. Meteorol. Sci. Technol.* **2011**, *39*, 227–230. (In Chinese) [CrossRef]
39. Ren, Z.; Feng, M.; Zhang, H.; Ju, X.; Wang, Y. Differences and correlations between automatic and manually observed rainfall. *Chin. J. Appl. Meteorol. Sci.* **2007**, *18*, 358–364. (In Chinese)
40. Ren, Z.; Zhao, P.; Zhang, Q.; Zhang, Z.; Cao, L.; Yang, Y.; Zouu, F.; Zhao, Y.; Zhao, H.; Chen, Z. Quality control methods for hourly precipitation data from automatic stations nationwide. *Chin. J. Meteorol. Mon.* **2010**, *36*, 123–132. (In Chinese)
41. Wang, X.; Xiao, P.; Feng, X.; Li, H. Large-scale urban land use information extraction in China based on DMSP/OLS data. *Chin. J. Remote Sens. Nat. Resour.* **2013**, *25*, 159–164. (In Chinese)
42. Pinto, O.; Pinto, I.R.C.A.; Ferro, M.A.S. A study of the long-term variability of thunderstorm days in southeast Brazil. *J. Geophys. Res. Atmos.* **2013**, *118*, 5231–5246. [CrossRef]
43. Miao, S.; Chen, F.; Li, Q.; Fan, S. Impacts of Urban Processes and Urbanization on Summer Precipitation: A Case Study of Heavy Rainfall in Beijing on 1 August 2006. *J. Appl. Meteorol. Climatol.* **2011**, *50*, 806–825. [CrossRef]
44. Ashley, W.S.; Bentley, M.L.; Stallins, J.A. Urban-induced thunderstorm modification in the Southeast United States. *Clim. Chang.* **2011**, *113*, 481–498. [CrossRef]
45. Paliaga, G.; Faccini, F.; Luino, F.; Turconi, L. A spatial multicriteria prioritizing approach for geo-hydrological risk mitigation planning in small and densely urbanized Mediterranean basins. *Nat. Hazards Earth Syst. Sci.* **2019**, *19*, 53–69. [CrossRef]
46. Jia, W.; Ren, G.; Tysa, S.K.; Zhang, P.; Wen, K.; Ren, Y. Urban heat island effect and its contribution to observed temperature increase at Wuhan station, Central China. *J. Trop. Meteorol.* **2019**, *25*, 102–113. [CrossRef]
47. Ren, G.Y.; Chu, Z.Y.; Chen, Z.H.; Ren, Y.Y. Implications of temporal change in urban heat island intensity observed at Beijing and Wuhan stations. *Geophys. Res. Lett.* **2007**, *34*, L05711. [CrossRef]
48. Shen, H.; Huang, L.; Zhang, L.; Wu, P.; Zeng, C. Long-term and fine-scale satellite monitoring of the urban heat island effect by the fusion of multi-temporal and multi-sensor remote sensed data: A 26-year case study of the city of Wuhan in China. *Remote Sens. Environ.* **2016**, *172*, 109–125. [CrossRef]

49. Wu, H.; Ye, L.-P.; Shi, W.-Z.; Clarke, K.C. Assessing the effects of land use spatial structure on urban heat islands using HJ-1B remote sensing imagery in Wuhan, China. *Int. J. Appl. Earth Obs. Geoinf.* **2014**, *32*, 67–78. [CrossRef]
50. Tysa, S.K.; Ren, G.; Qin, Y.; Zhang, P.; Ren, Y.; Jia, W.; Wen, K. Urbanization Effect in Regional Temperature Series Based on a Remote Sensing Classification Scheme of Stations. *J. Geophys. Res. Atmos.* **2019**, *124*, 10646–10661. [CrossRef]
51. Molina-Sanchis, I.; Lázaro, R.; Arnau-Rosalén, E.; Calvo-Cases, A. Rainfall timing and runoff: The influence of the criterion for rain event separation. *J. Hydrol. Hydromech.* **2016**, *64*, 226–236. [CrossRef]
52. Cao, S.; Diao, Y.; Wang, J.; Liu, Y.; Raimondi, A.; Wang, J. KDE-Based Rainfall Event Separation and Characterization. *Water* **2023**, *15*, 580. [CrossRef]
53. Fu, P. Diurnal Cycle of Rainfall over Yangtze and Huai River Basin during Meiyu and Analysis of a Typical Convective System in the Morning. Master's Thesis, Nanjing University, Nanjing, China, 2016.
54. Jiang, X.; Luo, Y.; Zhang, D.-L.; Wu, M. Urbanization Enhanced Summertime Extreme Hourly Precipitation over the Yangtze River Delta. *J. Clim.* **2020**, *33*, 5809–5826. [CrossRef]
55. Fu, X.; Yang, X.Q.; Sun, X. Spatial and Diurnal Variations of Summer Hourly Rainfall Over Three Super City Clusters in Eastern China and Their Possible Link to the Urbanization. *J. Geophys. Res. Atmos.* **2019**, *124*, 5445–5462. [CrossRef]
56. Li, M.; Wang, T.; Xie, M.; Zhuang, B.; Li, S.; Han, Y.; Cheng, N. Modeling of urban heat island and its impacts on thermal circulations in the Beijing–Tianjin–Hebei region, China. *Theor. Appl. Climatol.* **2016**, *128*, 999–1013. [CrossRef]

Disclaimer/Publisher's Note: The statements, opinions and data contained in all publications are solely those of the individual author(s) and contributor(s) and not of MDPI and/or the editor(s). MDPI and/or the editor(s) disclaim responsibility for any injury to people or property resulting from any ideas, methods, instructions or products referred to in the content.

land

MDPI

Article

Impacts of Urbanization and Its Parameters on Thermal and Dynamic Fields in Hangzhou: A Sensitivity Study Using the Weather Research and Forecasting Urban Model

Mengwen Wu [1,2], Meiying Dong [1], Feng Chen [1,*] and Xuchao Yang [3]

[1] Zhejiang Institute of Meteorological Sciences, Hangzhou 310008, China; wmw412@163.com (M.W.); dongmy_zj@163.com (M.D.)
[2] State Key Laboratory of Severe Weather, Chinese Academy of Meteorological Sciences, Beijing 100081, China
[3] Ocean College, Zhejiang University, Zhoushan 316021, China; yangxuchao@zju.edu.cn
* Correspondence: fchen_zj@163.com

Abstract: The impact of urbanization and the sensitivity of urban canopy parameters (UCPs) on a typical summer rainfall event in Hangzhou, China, is investigated using three groups of ensemble experiments. In this case, urbanization leads to higher temperatures, lower mixing ratios, lower wind speeds before precipitation, and more precipitation in and around the urban area. Both the thermal and dynamical effects of urbanization contribute to an increase in temperature and precipitation, with thermal effects contributing 71.2% and 63.8% to the temperature and precipitation increase, respectively, while the thermal and dynamical impacts cause the opposite changes to the mixing ratio and wind speed. Compared to the other three meteorological elements, the model has the largest uncertainty in the simulation of precipitation, which includes the sensitivity of the different parameterization schemes to the simulation of precipitation in urban areas, and the uncertainty brought by the urban effect on precipitation is not confined within the city but extends to the surrounding areas as well. Temperature and mixing ratio are more sensitive to thermal-related UCPs, while the wind speed is mainly affected by the structural parameters. These variations, however, are sometimes contradictory to precipitation changes, which further adds to the complexity of precipitation simulation.

Keywords: LULC; urban canopy parameter; local heavy precipitation; uncertainty

1. Introduction

Land cover and land use (LCLU) changes disrupt surface energy fluxes and thus interfere with exchanges of heat, water, and aerosols between the land surface and the atmosphere [1]. Of all the underlying surface types, large cities are experiencing the highest intensity of LCLU changes, which leads to a series of climate and environmental issues, such as urban heat islands (UHIs) [2–4], extreme rainfall [5,6], and air pollution [7,8]. With rapid urbanization occurring since the late last century, many observational studies show statistically significant increases in the frequency of short-duration extreme rainfall in major megacities during the warm season [9,10]. In addition, most of the extreme rainfall events take place under the influences of warm and wet airstreams with weak baroclinicity [11–13].

In the last few years, our understanding of the impact of urbanization has greatly advanced thanks to Numerical Weather Prediction (NWP) models [14–16]. To consider the effects of urban areas within NWP models, different parameterizations have been developed, for example, the single-layer urban canopy model (SLUCM) [17,18] and building effect parameterization (BEP) [19], to take into account urban-scale processes within mesoscale models, and they are currently implemented in the Weather Research and Forecasting (WRF) model [20]. In general, urbanization can produce less soil moisture, higher surface temperature, and lower average wind speed [16,21,22]. However, the urban modification of rainfall and its dominant physical mechanisms is more complex [23–25]. For example, the UHI magnitude will determine different impacts on precipitation enhancements spatially [26,27]. The strong UHI will increase atmospheric instability in the boundary layer and help transport more water vapor into the upper atmosphere, resulting in both an increased amount and frequency of extreme precipitation over cities [11,28,29]. In contrast, in the presence of strong winds or weak UHI, convective systems usually bifurcate over cities due to the building barrier effect, resulting in reduced rainfall in both urban centers and downwind areas [30,31]. More studies agree that the impact of urban land cover on precipitation is not only determined by their influence on local convergence but also modulated by large-scale weather systems [32]. For cases with a strong synoptic forcing, the urban impact is relatively weak [33]. When the synoptic forcing is weak, however, the urban impact on local convergence dominates [31]. Therefore, the feedback mechanism of cities to rainfall is highly uncertain.

To correctly perform numerical simulations for assessing the role of cities, an adequate representation of urban canopy parameters (UCPs) is required [34–36]. In fact, the urban morphology determined by UCPs can greatly affect the thermal and dynamical fields in urban areas [37]. For instance, the increased surface roughness due to the construction of buildings plays a positive role in convective initiation/enhancement [38], evapotranspiration decreases by changing vegetated surfaces into impervious surfaces [39], and the decreased surface albedo, increased building height, and increased anthropogenic heat release are sensitive to surface temperature [40,41] and UHI intensity [42]. These altered environmental factors can further influence when, where, where, and how intense urban precipitation occurs [43]. Some studies further suggest that the relationship between model performances and UCPs varied in different weather conditions and meteorological elements. For example, the impacts of updating urban fraction and urban morphology were obvious on wind speed but minor on temperature and humidity in both Beijing and Guangzhou, China [44,45]. The LCLU changes due to urban expansion lead to a reduction in precipitation in the Beijing–Tianjin–Hebei metropolitan region [46,47] but an increase in the Yangtze River Delta [48,49] and the Guangdong–Hong Kong–Macao Greater Bay Area [50,51]. However, these studies mainly focus on the impact of changes in either a single UCP or the entire UCP dataset on the urban environment; a diagnostic analysis of isolating the contributions of these UCPs on the thermal and dynamic fields has been conducted fewer times.

To better explain why the modeling of precipitation in urban areas has large uncertainties and how urbanization affects precipitation, three sets of sensitivity simulations, including changes in parameterization schemes, LCLU, and UCPs, are carried out to study a typical summer heavy rainfall event over the city of Hangzhou, China. Hangzhou had an urban (impervious) surface area of around 8289 km^2 and a permanent population of 12.3 million in 2022 (Zhejiang Statistical Bureau; http://tjj.zj.gov.cn/col/col1525563/index.html accessed on 20 July 2023). Wide roads and high-rise buildings are being constructed in Hangzhou, which changes the physical properties of the underlying surface and has a great impact on the urban environment [52]. Located in the Yangtze River Delta adjacent to the coast, Hangzhou, like many other major cities on the east coast of China, is quite vulnerable to local heavy precipitation that occurs abruptly. These factors make the city ideal for investigating the urban mechanisms responsible for local heavy rainfall modification. The paper is organized as follows: The description of the synoptic background, model configuration, and sensitivity experiments are given in Section 2. Section 3 evaluates the performance of the model and discusses the impacts of urbanization and UCPs on this rainfall event. The discussion and conclusion are presented in Section 4 and Section 5, respectively.

2. Materials and Methods

2.1. Synoptic Background

The selected summer rainfall case occurred between 13:00 and 18:00 Beijing Standard Time (BST; +8 h Coordinated Universal Time) on 26 July 2018, with a maximum hourly rainfall amount exceeding 114.7 mm at Liancheng Station (30.367° N, 120.283° E) near downtown Hangzhou. Before and during this event, Hangzhou was dominated by the uniform West Pacific subtropical high-pressure system, with relatively weak winds and less favorable moisture conditions (Figure 1). Moreover, high-temperature phenomena (≥38 °C at 10:00 BST) were observed in Hangzhou before the precipitation. The UHI intensity reached up to 0.77 °C during 10–13 BST, which is larger than the historical 10-year average (2012–2021) UHI intensity (0.48 °C) of Hangzhou in the same period in summer. This synoptic pattern is not conducive to forming regional rainfall but is favorable to initiating local rainfall. Therefore, this case is suitable to study the impacts of urbanization on local heavy precipitation and the prior environmental conditions.

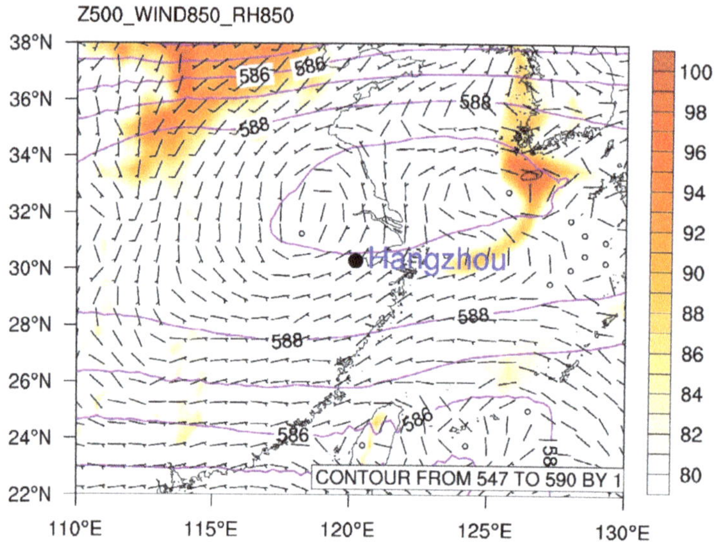

Figure 1. The ERA5 reanalysis of geopotential height at 500 hPa (solid lines), wind speed (arrow), and relative humidity (shading) at 850 hPa averaged from 09:00 to 18:00 on 26 July 2018. The black dot indicates the location of Hangzhou.

2.2. Model Configuration

The model used in the current study is the WRF-ARW coupled with urban model version 4.0.2 (http://www2.mmm.ucar.edu/wrf/users accessed on 1 May 2021), which is widely used for mesoscale numerical simulation research around the world [53]. Figure 2a shows the terrain of our study area and the location of Hangzhou. All the simulations have two one-way nested domains centered at (31.0° N, 120.5° E), with a horizontal grid spacing (grid numbers) of 4 km (353 × 403) and 1 km (457 × 413), respectively. The vertical coordinate contains 51 full-sigma levels from the surface to 10 hPa. Approximately eight of these levels are designed below 1 km to provide a fine vertical resolution within the planetary boundary layer, and the lowest half-sigma level height is approximately 35 m above ground. The black rectangle box (119.9–120.8° E, 29.8–30.7° N) in Figure 2b is referred to as the control region in this study, as both the urban area and heavy precipitation are concentrated in this region.

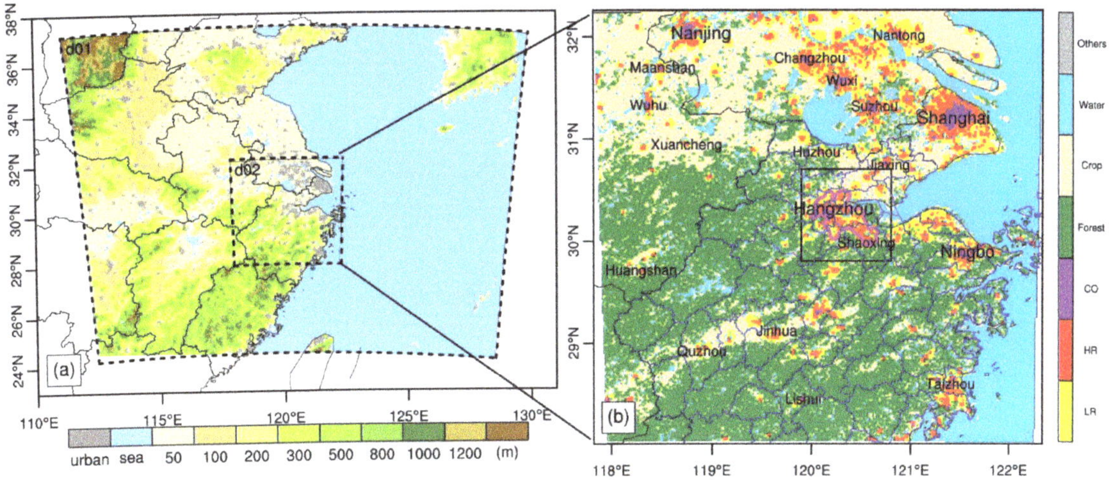

Figure 2. The (**a**) simulated domain d01 with terrain (m) and (**b**) land use in domain d02. The black rectangle box (119.9–120.8° E, 29.8–30.7° N) represents the control region of this heavy rain event. The magenta solid line in (**b**) indicates the Hangzhou–Shaoxing mega-city with an urban fraction ≥0.2. The LR, HR, and CO in (**b**) represent the land use type of low-density residential, high-density residential, and commercial, respectively.

The description of the model configuration is summarized in Table 1. Considering that the ensemble simulation is a more effective way of assessing the urban impact on local weather [54,55], 24 tests were designed using three microphysical schemes, two planetary boundary layer parameterization schemes, two land surface schemes, and two urban surface schemes, respectively. In addition, the RRTMG [56] is employed for both shortwave and longwave radiation schemes in the simulation. No cumulus parameterization is used, given that the model can resolve deep convective updrafts [53]. All the physical schemes used are identical for both nests.

Table 1. Model configurations.

Model Settings	D01	D02
Model and version	WRF v4.0.2	
Horizontal grid points	403 × 353	413 × 457
Δx (km)	4	1
Vertical layers	51	
Cumulus physics	None (0) *	
Shortwave radiation	RRTMG (4) [56]	
Longwave radiation	RRTMG (4) [56]	
Microphysics	Purdue Lin (2) [57], WSM6 (6) [58], WDM6 (16) [59]	
PBL physics	MYJ (2) [60,61], Boulac (8) [62]	
Land surface	Noah (2) [63], NoahMP (4) [64]	
Urban physics	SLUCM (1) [24], BEP (2) [19]	

* The numbers in parentheses represent the option number for each physical parameterization scheme.

All the simulations were initialized at 02 BST on 26 July 2018, and integrated for 18 h. The hourly ERA5 reanalysis data [65] with a resolution of 0.25° × 0.25° were used to provide the initial and lateral boundary conditions. No data assimilation or nudging were performed to avoid the influence of other factors. Considering the rapid urban expansion in and around Hangzhou, the default MODIS 20 category land use dataset [66] used in our study was updated by a more inhomogeneous and detailed urban land cover and fraction dataset at 1 km × 1 km resolution [67]. Using this dataset has been proved to better simulate the UHIs [52,67] and extreme rainfall [68] in Hangzhou. The updated land use map is shown in Figure 2b, and the Hangzhou–Shaoxing mega-city with an urban fraction ≥0.2 is highlighted with a magenta solid line.

2.3. Numerical Experiment Design

Three groups of numerical experiments, i.e., GROUP I, II, and III, are designed in our study (Table 2). In GROUP I, there are a total of 24 ensemble members with the same model configurations, except for different combinations of physical parameterization schemes. Every ensemble member is named by the option of the selected physical parameterization schemes, e.g., the name of "m2p2s2u1" represents that Lin (2), MYJ (2), Noah (2), and SLUCM (1) are selected for the microphysics, PBL physics, land surface, and urban physics schemes of this ensemble member, respectively. The control ensemble forecast (ENCTL) includes these 24 members running with the updated real land use data. Another ensemble forecast (referred to as ENNoUB) is designed the same as the ENCTL. However, we replace the urban land use with cropland in each member to investigate the impact of urbanization.

Table 2. Summary of the numerical experiments.

Group	WRF Simulations	Physical Parameterization Options	Notes
GROUP I	ENCTL	Microphysics (Lin(2), WSM6(6), WDM6(16)) Planetary Boundary layer (MYJ(2), BouLac(8)) Land Surface (Noah (2), NoahMP(4)) Urban Surface (SLUCM(1), BEP(2))	1. Choose one option from each physical group, and make up a member named by the option numbers, e.g., m2p2s2u1 represents the Lin(2), MYJ(2), Noah(2), and SLUCM(1) schemes used in this member. 2. 24 members are chosen for the numerical simulation, and the m16p8s4u2 is chosen as the control run (CTL) because of its good performance. 3. The real terrain and land use data are used in the numerical simulation (Figure 2).
	ENNoUB	Same as ENCTL	Same as ENCT, replace the urban land use type with cropland.
GROUP II	CTL	WDM6(16) + BouLac(8) + Noahmp(4) + BEP(2)	The control run in ENCTL, abbreviated as m16p8s4u2 run.
	NOUB	Same as CTL	Same as CT but artificially remove both the thermal and dynamical effect of urbanization.
	NOTH	Same as CTL	Same as CT but artificially removes the thermal effect of urbanization.
	NODY	Same as CTL	Same as CTL but artificially removes the dynamical effect of urbanization.
GROUP III	CTL	WDM6(16) + BouLac(8) + Noahmp(4) + BEP(2)	The control run in ENCTL, abbreviated as m16p8s4u2 run.
	SEN1/SEN2	Same as CTL	Same as CTL but decreases/increases the building height by 50%.
	SEN3/SEN4	Same as CTL	Same as CTL but decreases/increases the roof width by 50%.
	SEN5/SEN6	Same as CTL	Same as CTL but decreases/increases the road width by 50%.
	SEN7/SEN8	Same as CTL	Same as CTL but decreases/increases the anthropogenic heat by 50%.
	SEN9/SEN10	Same as CTL	Same as CTL but decreases/increases the heat capacity by 50%.
	SEN11/SEN12	Same as CTL	Same as CTL but decreases/increases the thermal conductivity by 50%.
	SEN13/SEN14	Same as CTL	Same as CTL but decreases/increases the surface albedo by 50%.
	SEN15/SEN16	Same as CTL	Same as CTL but decreases/increases the roughness length by 50%.

In GROUP II, three sensitivity tests (referred to as NOTH, NODY, and NOUB) are conducted and compared to the control simulation called CTL (the member of "m16p8s4u2" in ENCTL) to investigate the thermal and dynamical effects of urbanization on this event. The NOTH, NODY, and NOUB runs are designed the same as the CTL, but we artificially removed the urban thermal and dynamical effects, respectively. The urban process, as a part of the land surface process, influences the bottom boundary variables in WRF following Equation (1).

$$V_{grid} = V_{veg} \times (1 - w_{urb}) + V_{urb} \times w_{urb} \tag{1}$$

where V_{grid} is the variable (e.g., surface temperature) from the surface to the lowest layer of the grid; V_{veg} is the same variable from the land surface model for the natural vegetation surfaces; V_{urb} is the same variable from the urban canopy model for the artificial surfaces; and w_{urb} is the fraction coverage of artificial surfaces. The urban thermal effect is removed by setting $w_{urb} = 0$ for the thermal-related variables, i.e., albedo, surface temperature (TSK), sensible heat flux (HFX), latent heat flux (LH), and ground heat flux (GRDFLX). The dynamical effect is removed by setting $w_{urb} = 0$ for the dynamical-related variables, i.e., friction velocity (UST) and momentum.

In GROUP III, another 16 sensitivity tests (i.e., SEN1, SEN2, ..., and SEN16) are conducted. Each sensitivity test is the same as the CTL but with a decrease or increase in one UCP (such as the building height) by 50%. The sensitivity results are compared with the CTL to systematically explore the effects of UCPs on the convective initiation and the prior urban environmental conditions.

The observed meteorological variables, including precipitation, 2 m temperature (T2M), 2 m mixing ratio (Q2M), and 10 m wind speed (WS10M), are derived from the minute-level surface observations maintained by the Zhejiang Meteorological Bureau, China Meteorological Administration. The gauged variables are interpolated into the model grid using the inverse distance-weighted method [69,70]. Thus, the gridded observation data are used for a quantitative evaluation of the model performance. The evaluation of the effect of urbanization on this precipitation event is divided into three components: (1) on the atmospheric conditions (e.g., T2M, Q2M, and WS10M) prior to the rainfall occurrence (from 10:00 to 13:00 BST on 26 July 2018); (2) on the accumulated precipitation (from 13:00 to 18:00 BST on 26 July 2018); and (3) on convective initiation by changes in the UCPs (from 13:00 to 14:00 BST on 26 July 2018).

3. Results

3.1. Evaluation of WRF Simulation

The accumulated precipitation from 13:00 to 18:00 BST on 26 July 2018 is simulated using 24 ensemble tests with different combinations of physical schemes using a map of current land use (ENCTL in Group I). The threat score (*TS*) is used to evaluate the performance of each test in predicting the accumulated precipitation (Table 3). The *TS* is defined as follows:

$$TS = \frac{N_a}{N_t - N_b} \tag{2}$$

where N_a is the number of simulations that capture the observation, N_b is the number of missing grids, and N_t is the number of total grids, respectively.

The performance of the model in capturing the occurrence of rainfall (≥0.1 mm) is generally good, with an average of $TS > 0.85$ accuracy, although the model performance gradually decreases at higher accumulations. Comparing the *TS* among all the members in each rainfall intensity category (≥0.1, ≥10, ≥25, and ≥50 mm), the microphysics scheme plays a key part in forecasting rainfall, and the majority of the members with higher *TS* use the microphysics scheme of WDM6 (16). When other parametrization schemes are the same, members using the BEP scheme usually obtain higher *TS*s than those using SLUCM. Yu et al. [44] also suggest that the BEP scheme performs better than the SLUCM scheme over the urban core of Beijing. A combination of the WDM6 (16), BouLac (8), Noah MP (4), and BEP (2), i.e., m16p8s4u2, finds the highest *TS* when precipitation is ≥10 and ≥25 mm and the second-highest *TS* when precipitation is ≥0.1 mm. From the evolution of radar echoes, the convective system first initiates in the southeast corner of the control region at around 14:00 BST (Figure 3(a1)). Then, it moves northwestward with a rapid intensification in the next two hours (Figure 3(b1,b2)). The radar echo reaches its maximum value at 16:00 BST in downtown Hangzhou and then gradually weakens and moves out of the control region in a northwesterly direction (Figure 3(d1)). In general, the m16p8s4u2 successfully captures the evolution of the heavy-rain-producing convective system, except for an approximate shift of 1.5 h in advance (Figure 3(a2–d2)). Therefore, the m16p8s4u2 in the ENCTL is chosen as the CTL.

Table 3. Threat scores (*TS*s) of accumulated precipitation (from 13:00 to 18:00 BST on 26 July 2018) for the ensemble study on different physical schemes *.

CASES	*TS* (≥0.1 mm)	*TS* (≥10 mm)	*TS* (≥25 mm)	*TS* (≥50 mm)
m2p2s2u1	0.87	0.32	0.16	0.02
m2p2s2u2	0.92	0.47	0.27	0.07
m2p2s4u1	0.90	0.41	0.21	0.04
m2p2s4u2	0.93	0.52	0.24	0.07
m2p8s2u1	0.84	0.44	0.19	0.02
m2p8s2u2	0.87	0.43	0.19	0.03
m2p8s4u1	0.82	0.43	0.28	0.16
m2p8s4u2	0.86	0.41	0.19	0.06
m6p2s2u1	0.87	0.34	0.14	0.00
m6p2s2u2	0.88	0.34	0.07	0.00
m6p2s4u1	0.90	0.36	0.16	0.02
m6p2s4u2	0.85	0.37	0.14	0.02
m6p8s2u1	0.86	0.28	0.10	0.11
m6p8s2u2	0.92	0.39	0.14	0.02
m6p8s4u1	0.84	0.43	0.17	0.04
m6p8s4u2	0.84	0.40	0.12	0.00
m16p2s2u1	0.92	0.53	0.26	0.03
m16p2s2u2	0.92	0.48	0.15	0.02
m16p2s4u1	0.90	0.49	0.15	0.02
m16p2s4u2	0.92	0.61	0.34	0.02
m16p8s2u1	0.86	0.52	0.20	0.04
m16p8s2u2	0.93	0.53	0.19	0.02
m16p8s4u1	0.90	0.60	0.30	0.15
m16p8s4u2	0.92	0.56	0.33	0.07
Mean	0.87	0.37	0.14	0.03

* The colored grid points are the top 25% of members. The darker the color, the higher the model performance.

The averaged T2M, Q2M, and WS10M before precipitation (from 10:00 to 13:00 BST on 26 July 2018) and the accumulated precipitation (from 13:00 to 18:00 BST on 26 July 2018) of the ENCTL are both qualitatively (Figure 4) and quantitatively (Table 4) compared with the observations. The deviation (*BIAS*) and root-mean-square error (*RMSE*) based on the following definitions are used to analyze the averaged uncertainty of the four meteorological variables.

$$BIAS = \frac{1}{N}\sum_{i=1}^{N}(X_i - Y_i) \tag{3}$$

$$RMSE = \frac{1}{N}\sum_{i=1}^{N}(X_i - Y_i)^2 \tag{4}$$

where X_i and Y_i represent the variable in each grid from WRF model and observation, respectively.

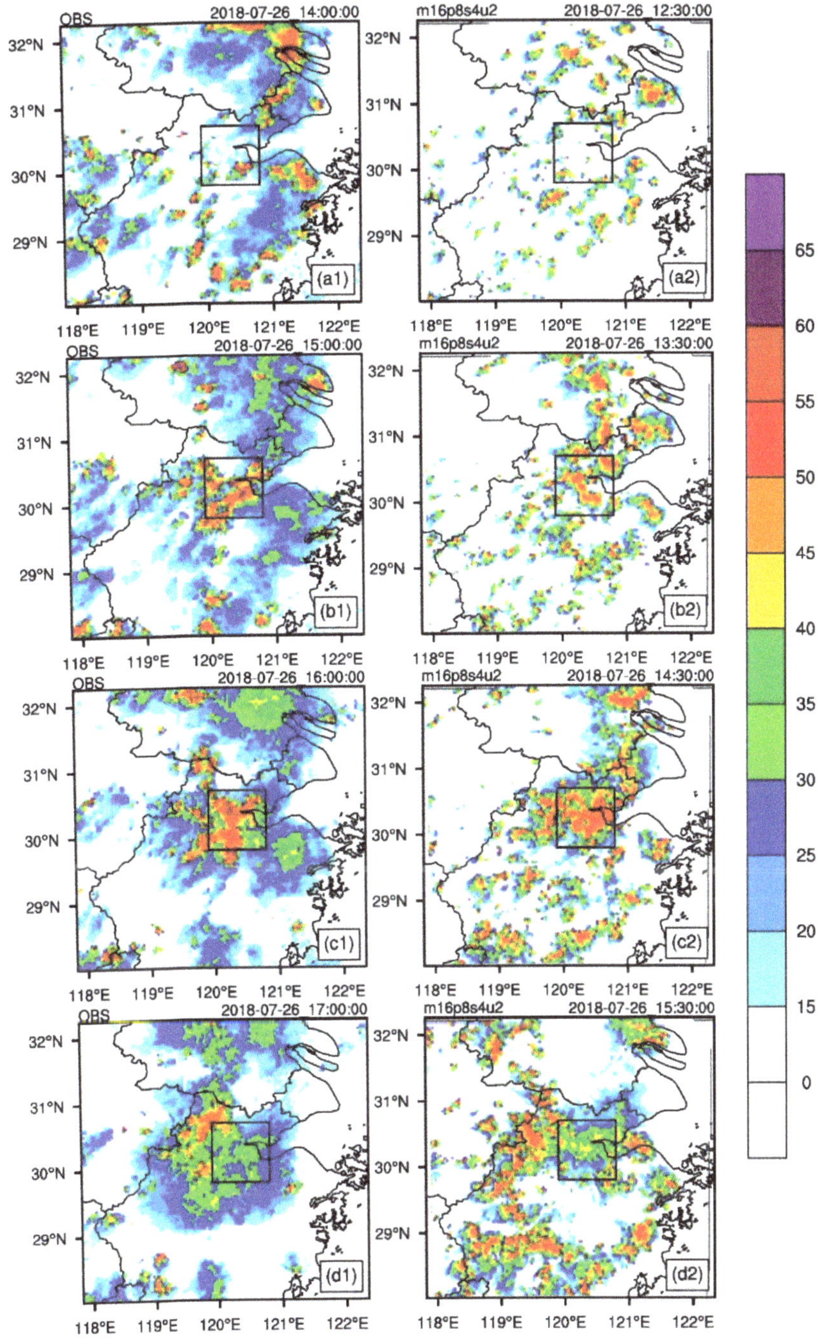

Figure 3. Comparison of the (**a1–d1**) observed and (**a2–d2**) simulated (the member of "m16p8s4u2" in ENCTL, CTL for short) composite radar reflectivity (dBZ) at four selected times on 26 July 2018. Note an approximate shift of 1.5 h between the simulation and observation.

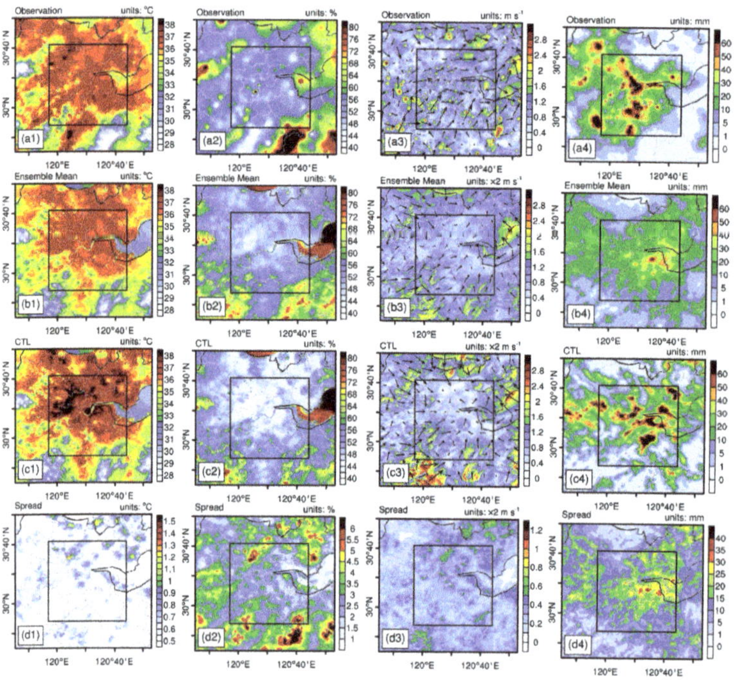

Figure 4. Distribution of the 2 m temperature (°C) averaged during 10:00 to 13:00 BST on 26 July 2018: (**a1**) observation, (**b1**) ensemble mean, (**c1**) CTL run, and (**d1**) spread; (**a2–d2**) the same as (**a1–d1**) but for the 2 m mixing ratio (g kg^{-1}); (**a3–d3**) the same as (**a1–d1**) but for the 10 m wind (m s^{-1}); (**a4–d4**) similar to (**a1–d1**) but for the accumulated precipitation (mm) from 13:00 to 18:00 BST on 26 July 2018.

Table 4. The deviation (*BIAS*) and root-mean-square error (*RMSE*) of accumulated precipitation (from 13:00 to 18:00 BST on 26 July 2018) for the ensemble study on different physical schemes.

Variables	Scores	Ensemble Mean	Ensemble Variation
T2M	*BIAS* (°C)	−1.43	0.27
	RMSE (°C)	1.97	0.16
Q2M	*BIAS* (g kg^{-1})	−1.83	0.42
	RMSE (g kg^{-1})	2.54	0.20
WS10M	*BIAS* (m s^{-1})	1.31	0.13
	RMSE (m s^{-1})	1.73	0.09
PREC	*BIAS* (mm)	−6.14	13.64
	RMSE (mm)	29.08	2.22

Compared with the observation (Figure 4(a1)), the ensemble means of ENCTL (Figure 4(b1)) capture the main features of T2M, with a slight underestimation in the southern part of the control region, and the CTL successfully reports an extremely high temperature of 39 °C in the urban area (Figure 4(c1)). The spread of T2M is small in the ENCTL (Figure 4(d1)), indicating that the high temperature and the UHI are both well reproduced by most members. Accompanied by the hot weather, Hangzhou is also characterized by low moisture (Figure 4(a2)) and low wind speed (Figure 4(a3)). Both the ENCTL and the CTL ensemble mean capture these features, i.e., relatively smaller values of Q2M (Figure 4(b2,c2)) and WS10M (Figure 4(b3,c3)) within the control region than those outside the control region. These scenarios, known as the urban heat islands, dry islands, and turbid islands (exacerbated by low wind speed), are similar to those shown by previous studies [19,36,54].

The accumulated precipitation is mainly concentrated in the control region (Figure 4(a4)). The CTL well reproduces the spatial distribution and magnitude of the accumulated precipitation (Figure 4(c4)). In the ensemble mean of the ENCTL, precipitation is also concentrated in the control region, although the rainfall magnitude is underestimated, and the rainfall center shifts about 35 km to the east (Figure 4(b4)). This is a common issue that signals that extreme rainfall may be weakened by the ensemble mean [55]. Different from the T2M, Q2M, and WS10M, the spread of precipitation (Figure 4(d4)) is much larger inside the control region, indicating more uncertainties in forecasting precipitation in the urban area than in natural land use.

The ensemble means of *BIAS* are −1.43 °C, −1.83 g kg^{-1}, 1.31 m s^{-1}, and −6.14 mm for T2M, Q2M, WS10M, and precipitation, respectively. The ensemble means of *RMSE* are 1.97 °C, 2.54 g kg^{-1}, 1.73 m s^{-1}, and 29.08 mm for T2M, Q2M, WS10M, and precipitation, respectively (Table 4). Compared to the magnitude of the ensemble means, the variations in *BIAS* and *RMSE* are relatively small in T2M, Q2M, and WS10M compared to those in precipitation, which suggests that the performance of the model in forecasting the first three variables could be improved less by changing the options of various physical schemes but quite valuable for forecasting short-term local heavy rainfall.

A comparison of the hourly variations in T2M, Q2M, WS10M, and precipitation averaged within the control region between the observation and the ENCTL suggests that most of the members capture the evolution of this event, especially the T2M variation before the convective initiation (Figure 5). However, the T2M drops faster during the precipitation period and remains at a higher level in the ENCTL than in the observation (Figure 5a). In addition, most of the members underestimate Q2M (Figure 5b) but overestimate WS10M (Figure 5c) during the study period. The wind speed overestimations are also found for other cities in summertime as well [71]. The premature drop in T2M may be related to the early triggering of convection. Furthermore, similar to the CTL (Figure 3), the precipitation starts 0–2 h earlier in most members (Figure 5d), which may be a common bias in simulating local convection.

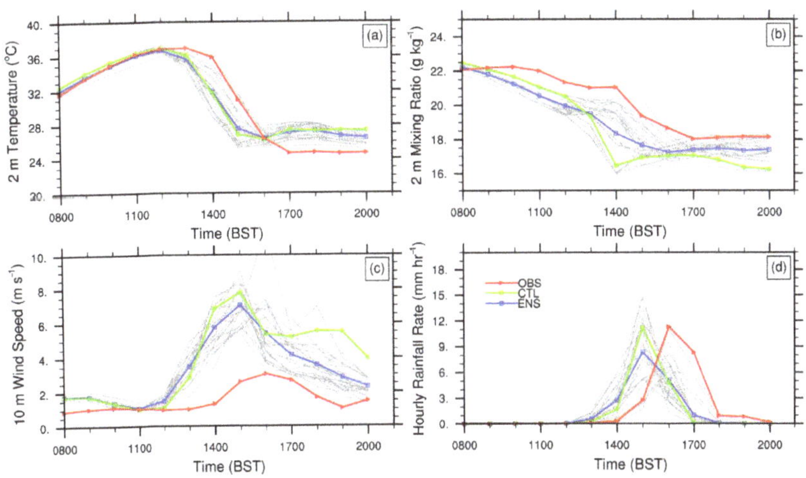

Figure 5. Time series of the (**a**) 2 m temperature (°C), (**b**) 2 m mixing ratio (g kg^{-1}), (**c**) 10 m wind speed (m s^{-1}), and (**d**) precipitation (mm) averaged over the control region from 08:00 to 20:00 BST on 26 July 2018. The red line denotes observation, the gray lines are the 24 ensemble members, the green line is the CTL (i.e., m16p8s4u2 in ENCTL), and the blue line represents the ensemble mean.

From the analysis above, the ENCTL captures the characteristics of meteorological elements before and during the heavy precipitation event. Compared with the environmental conditions (i.e., T2M, Q2M, and WS10M) before convective initiation, the ENCTL has greater uncertainty for simulating precipitation within the urban area. The ensemble member of m16p8s4u2 in the ENCTL is chosen as the control run (CTL) in this study. For the next two subsections, the individual and combined impacts of thermal and dynamical effects of urbanization and the impact of individual UCPs on this local heavy rainfall event are further examined.

3.2. Evaluation of the Urban Effect

The impact of urbanization is first examined by comparing the differences in T2M, Q2M, WS10M, and accumulated precipitation between ENCTL and ENNoUB (Figure 6). The increase in the ensemble mean of T2M is 0.4–1.2 °C in the urban area, which is much larger than in the surrounding nonurban areas (Figure 6a). On the contrary, urbanization causes the mixing ratio to decrease by 0.5–1.8 g kg^{-1} in the urban area (Figure 6c). This is consistent with previous findings that evaporation in the urban area is reduced by impervious surfaces and reduced green vegetation fraction [46,72]. The wind speed decreases by 0.3–1.0 m s^{-1} in the urban area due to the drag effect of buildings (i.e., the pressure differences across individual roughness elements; Figure 6e). The positive anomaly center of precipitation is located at the junction between the urban area and Hangzhou Bay (Figure 6g). The presence of urban areas by the bayside can enhance the thermodynamic contrast between land and water that favors the water–land breeze front and promotes the formation of a convergence zone [73,74]. Meanwhile, the differences in the spread of T2M (Figure 6b), Q2M (Figure 6d), and WS10M (Figure 6f) are mainly concentrated in the urban area; however, the difference in the spread of precipitation occurs both in and out of the control region (Figure 6h). This suggests that the uncertainty brought about by the urban effect on precipitation occurs within the city and affects the surrounding areas to a similar extent. Notably, the spatial distribution of spread in precipitation is similar to that of the ensemble mean (Figure 6g), which indicates that the ensemble method is quite valuable for assessing the urban impact on local heavy precipitation [75].

The individual and combined impacts of the thermal and dynamical effects of urbanization are further examined by comparing the simulated results of NUB, NTH, and NDY with CTL (GROUP II). Both the thermal and dynamical effects contribute to an increase in T2M (Figure 7(a1–a3)). The regional averaged values of T2M suggest that the contribution of the thermal effect is more than twice that of the dynamical effect (71.2% vs. 28.8%; Figure 8a). For changes in Q2M, although the dynamical effect is a positive contribution, the thermal effect plays a major role and contributes to a decrease in Q2M (Figure 7(b1–b3) and 8b). The thermal and dynamical effects are also opposite to the changes in wind speed in the urban area (Figure 7(c2) vs. Figure 7(c3)). However, since the dynamical effect is dominant (Figure 8c), the WS10M in the urban area is decreased due to urbanization. This is also consistent with the difference in WS10M between ENCTL and ENNoUB (Figure 6e). Different from the first three variables, both the influences of thermal and dynamical effects of urbanization on precipitation present a positive and negative distribution in the urban area (Figure 7(d1–d3)). The regional averaged values suggest that they play an overall positive role and contribute to an increase in precipitation by 63.8% and 36.2%, respectively (Figure 8d). For the thermal effect, the increased temperature and thermal turbulence activity (Figure S1) help intensify atmospheric instability and convective initiation/enhancement. For the dynamical effect, increased friction velocity and turbulent kinetic energy flux (TKEP; Figure S2) could increase the turbulent mixing strength and facilitate convergence, increasing the precipitation in turn.

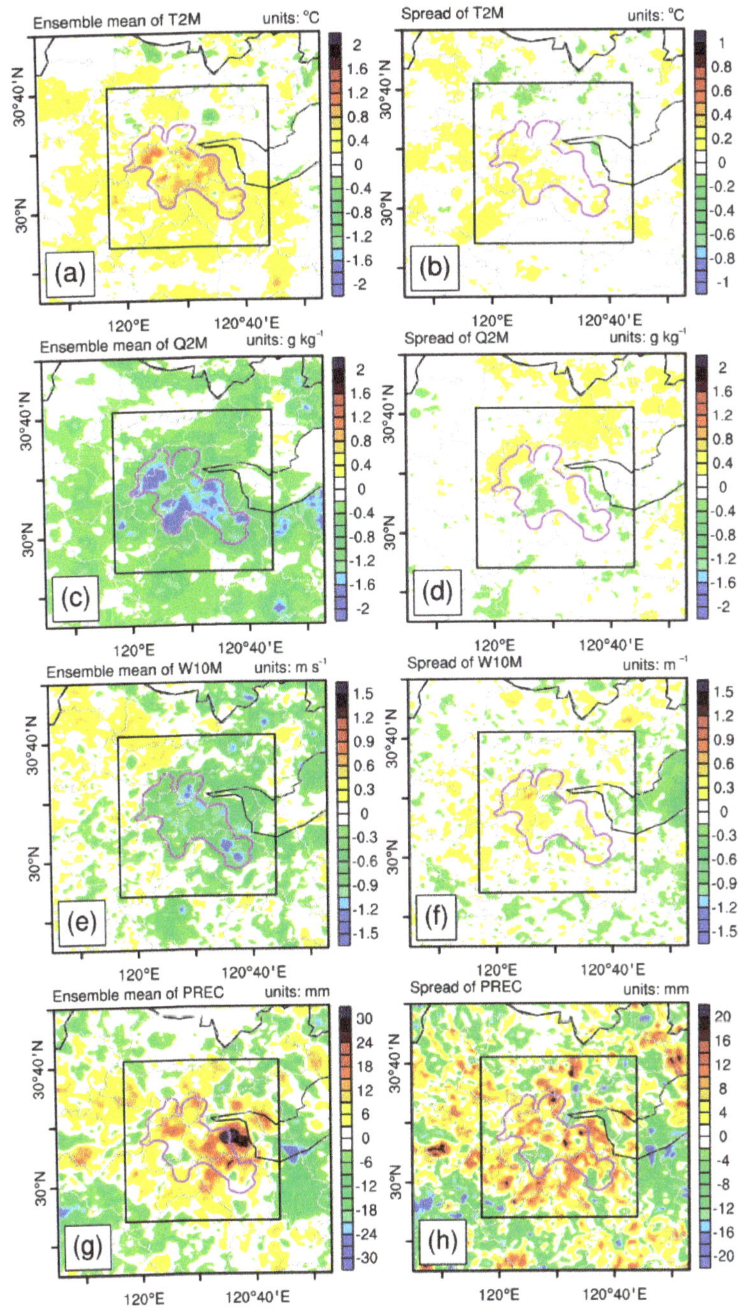

Figure 6. Differences between ENCTL and ENNoUB in the ensemble mean (**left**) and spread (**right**) of (**a**,**b**) 2 m temperature (°C) and (**c**,**d**) 2 m mixing ratio (g kg^{-1}), and (**e**,**f**) 10 m wind speed (m s^{-1}) during 10:00 to 13:00 BST on 26 July 2018, and (**g**,**h**) accumulated precipitation (mm) from 13:00 to 18:00 BST on 26 July 2018.

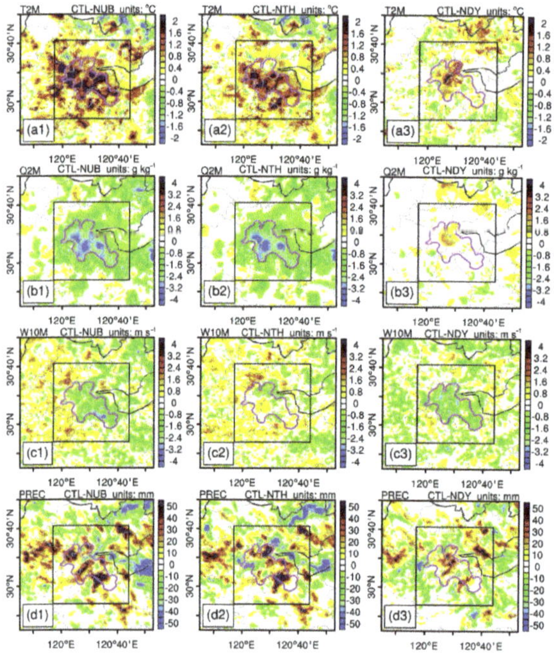

Figure 7. Differences in (**a1–a3**) 2 m temperature (°C) and (**b1–b3**) 2 m mixing ratio (g kg^{-1}), (**c1–c3**) 10 m wind speed (m s^{-1}) averaged during 10:00 to 13:00 BST on 26 July 2018, and (**d1–d3**) accumulated precipitation (mm) during 13:00 to 18:00 BST on 26 July 2018 between the control test (CTL) and (**left**) NoUB, (**middle**) NoTH, and (**right**) NoDY test.

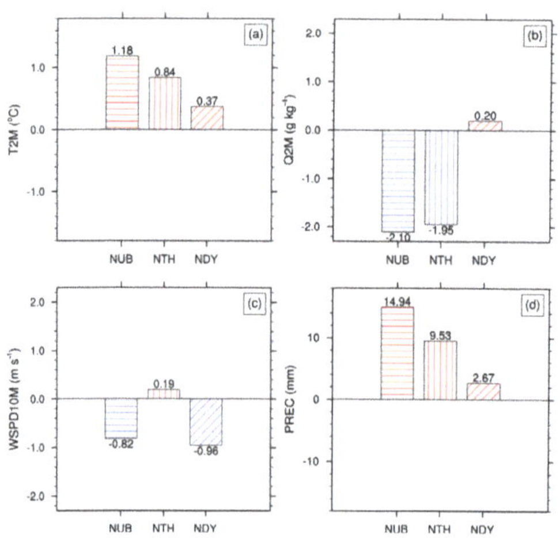

Figure 8. Comparison of the impacts of total, thermal, and dynamical effects of urbanization on (**a**) 2 m temperature (°C), (**b**) 2 m mixing ratio (g kg^{-1}), and (**c**) 10 m wind speed (m s^{-1}) averaged during 10:00 to 13:00 BST on 26 July 2018, and (**d**) accumulated precipitation (mm) averaged during 13:00 to 18:00 BST on 26 July 2018. The red column represents a positive contribution and the blue column represents a negative contribution.

3.3. Sensitivity Analysis of Urban Canopy Parameters

For such local, short-term heavy rainfall events, the environmental conditions before convective initiation are crucial for simulating the entire rainfall event. Most recent studies on urban climate and weather have viewed the role of UCPs as an integrated external force [45,76,77], but few of them mentioned the individual role of specific UCPs in the urban area. In this subsection, eight important parameters characterizing the urban canopy (i.e., building height, roof width, road width, anthropogenic heat, heat capacity, thermal conductivity, surface albedo, and roughness length) are selected and increased/decreased by 50%, separately (GROUP III). Each of the 16 sensitivity experiments is compared with the CTL for the differences in T2M (Figure 9), Q2M (Figure 10), and WS10M (Figure 11) before precipitation (at 10:00 BST on 26 July 2018) and of the subsequent convective initiation (during 13:00–14:00 BST on 26 July 2018; Figure 12), respectively.

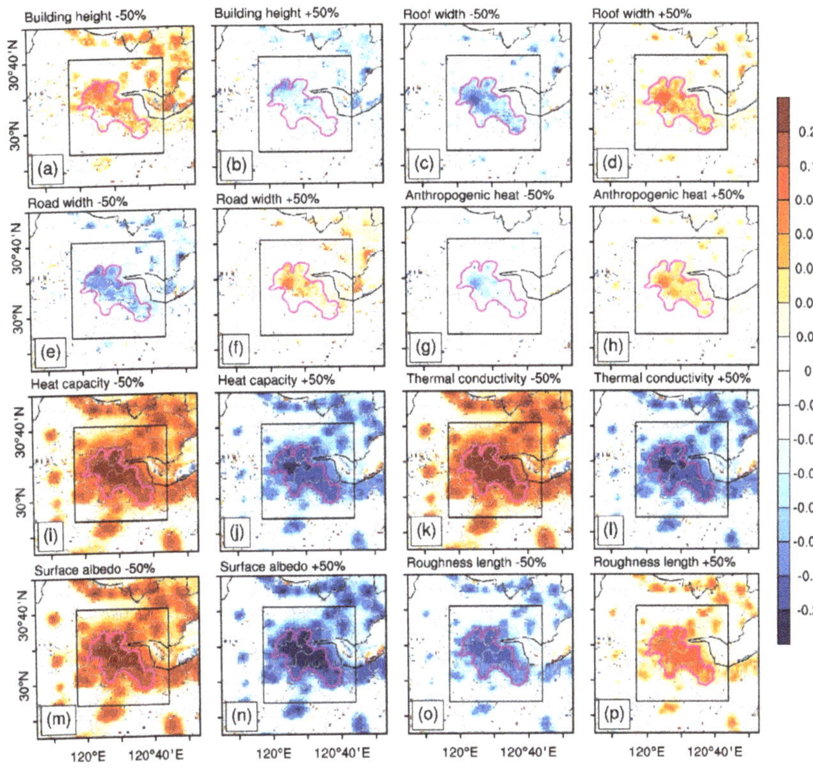

Figure 9. The temperature differences (°C) at 10:00 BST on 26 July 2018 between the 16 sensitivity tests in GROUP III (i.e., SEN1, SEN2, . . ., SEN16) and the CTL (m16p8s4u2) by (**a**) decreasing the building height; (**b**) increasing the building height; (**c**) decreasing the roof width; (**d**) increasing the roof width; (**e**) decreasing the road width; (**f**) increasing the road width; (**g**) decreasing the anthropogenic heat; (**h**) increasing the anthropogenic heat; (**i**) decreasing the heat capacity; (**j**) increasing the heat capacity; (**k**) decreasing the thermal conductivity; (**l**) increasing the thermal conductivity; (**m**) decreasing the surface albedo; (**n**) increasing the surface albedo; (**o**) decreasing the roughness length; and (**p**) increasing the roughness length by 50%.

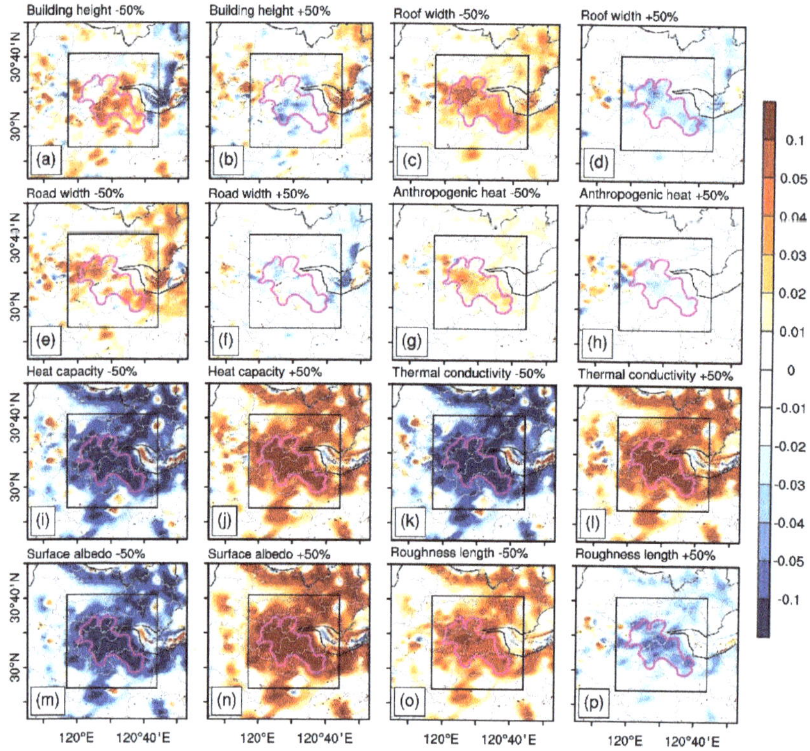

Figure 10. The same as Figure 9 but for the 2 m mixing ratio (g kg^{-1}): (**a**) decreasing the building height; (**b**) increasing the building height; (**c**) decreasing the roof width; (**d**) increasing the roof width; (**e**) decreasing the road width; (**f**) increasing the road width; (**g**) de-creasing the anthropogenic heat; (**h**) increasing the anthropogenic heat; (**i**) decreasing the heat capacity; (**j**) increasing the heat capacity; (**k**) decreasing the thermal conductivity; (**l**) increasing the thermal conductivity; (**m**) decreasing the surface albedo; (**n**) increasing the surface albedo; (**o**) decreasing the roughness length; and (**p**) increasing the roughness length by 50%.

For changes in T2M (Figure 9), UCPs of building height, heat capacity, thermal conductivity, and surface albedo have a negative feedback effect. The increased building height may mitigate the UHI intensity at the pedestrian level due to the shading effects of high-rise buildings [42]. Compared to the building height, T2M is more sensitive to the changes in the latter three thermal-related UCPs (Figure 9i–n). The temperature decreases by about 0.05–0.2 °C in the urban area by increasing the heat capacity, thermal conductivity, or surface albedo, and the changes in T2M are not limited to urban areas but reflected around the city as well. On the contrary, changes in roof width, road width, anthropogenic heat, and roughness length have a positive effect on T2M. Among these four UCPs, the roughness length has the most significant effect, increasing the temperature in urban areas by 0.04–0.08 °C (Figure 9o,p). The effects of UCPs on Q2M (Figure 10) are the opposite of those on T2M, except for changes in the building height. The seemingly contradictory contributions of increasing temperature and decreasing humidity to changes in precipitation make the uncertainty of the impact of urbanization on precipitation even greater.

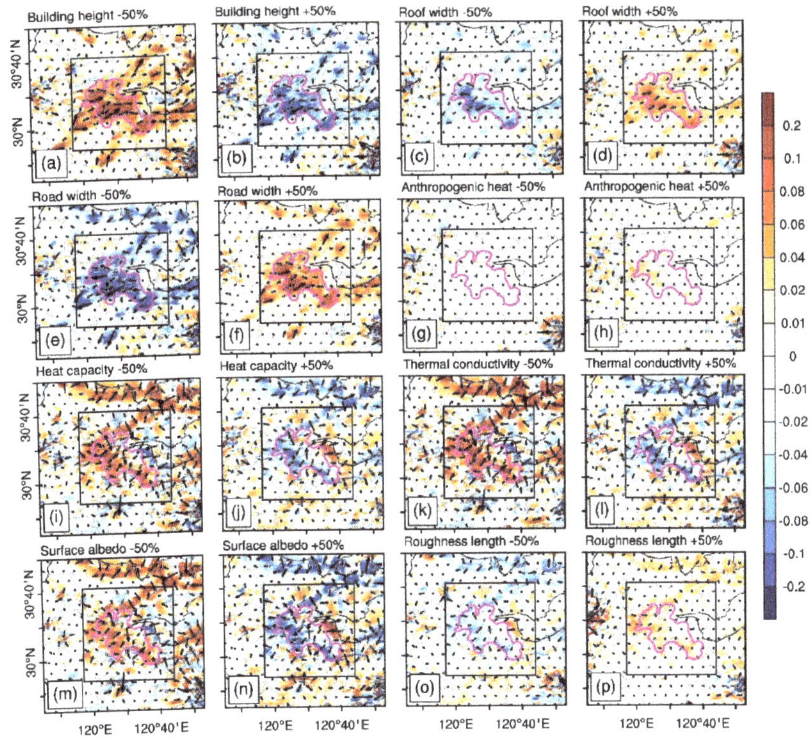

Figure 11. The same as Figure 9 but for the 10 m wind speed (m s^{-1}): (**a**) decreasing the building height; (**b**) increasing the building height; (**c**) decreasing the roof width; (**d**) increasing the roof width; (**e**) decreasing the road width; (**f**) increasing the road width; (**g**) de-creasing the anthropogenic heat; (**h**) increasing the anthropogenic heat; (**i**) decreasing the heat capacity; (**j**) increasing the heat capacity; (**k**) decreasing the thermal conductivity; (**l**) increasing the thermal conductivity; (**m**) decreasing the surface albedo; (**n**) increasing the surface albedo; (**o**) decreasing the roughness length; and (**p**) increasing the roughness length by 50%.

For the three structural parameters related to the changes in WS10M, the building height has a negative effect (Figure 11a,b), while the roof width (Figure 11c,d) and the road width (Figure 11e,f) are positive. In addition, three other thermal-related UCPs (i.e., heat capacity, thermal conductivity, and surface albedo) also have similar negative effects on WS10M, except at the junction of the city and the waters of Hangzhou Bay (Figure 11i–n). The decreased wind speed within the city and the increased wind speed at the edge of the city favor wind convergence and the subsequent convective initiation/enhancement in the urban area. This is consistent with the differences in the ensemble mean of the accumulated precipitation amount between ENCTL and ENNoUB (Figure 6g).

The influences of UCPs on precipitation during the convective initiation stage (Figure 12) are less uniformly distributed than those on the other three meteorological elements. Both positive and negative feedback effects can be observed in the urban area, suggesting large uncertainties in forecasting heavy precipitation there. This may be because, in response to favorable conditions for convective initiation, the UCPs sometimes have a contradictory effect on the urban environmental conditions (T2M, Q2M, and WS10M). For example, increasing the surface albedo favors moisture and wind convergence within the city but decreases the temperature and thermal instability, presenting an overall negative effect on convective initiation (Figure 12n). Once again, this illustrates the complexity of urban effects on precipitation.

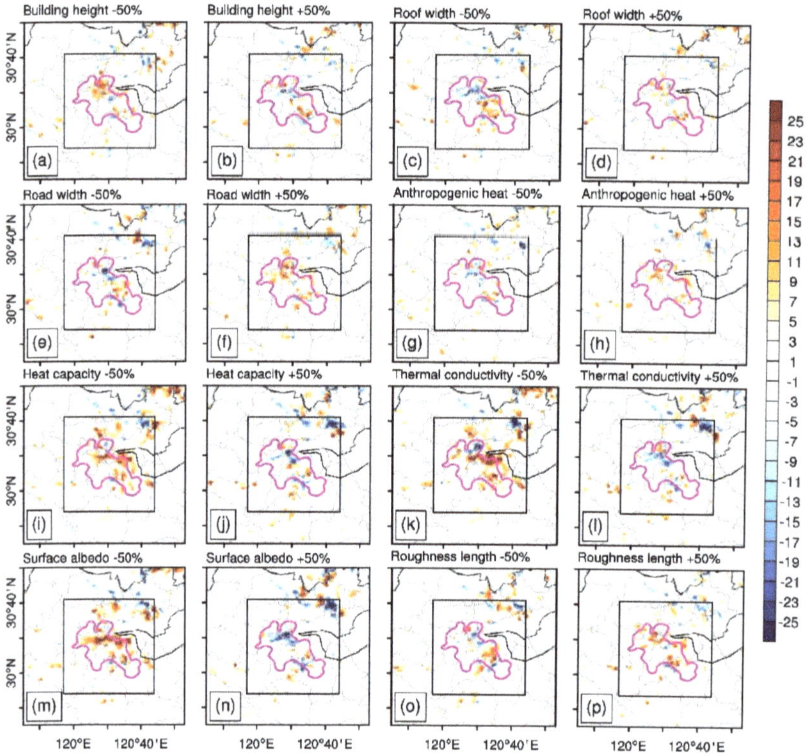

Figure 12. The same as Figure 9 but for 1 h accumulated precipitation (mm) during 13:00–14:00 BST on 26 July 2018: (**a**) decreasing the building height; (**b**) increasing the building height; (**c**) decreasing the roof width; (**d**) increasing the roof width; (**e**) decreasing the road width; (**f**) increasing the road width; (**g**) de-creasing the anthropogenic heat; (**h**) increasing the anthropogenic heat; (**i**) decreasing the heat capacity; (**j**) increasing the heat capacity; (**k**) decreasing the thermal conductivity; (**l**) increasing the thermal conductivity; (**m**) decreasing the surface albedo; (**n**) increasing the surface albedo; (**o**) decreasing the roughness length; and (**p**) increasing the roughness length by 50%.

Figure 13 lists the differences in more meteorological variables (*X*-axis) averaged in the control region when increasing the values of the eight UCPs (*Y*-axis) by 50% individually to better illustrate the changes in the environmental conditions within urban areas. The surface albedo and the three structural parameters (i.e., building height, roof width, and road width) play an important role in the urban canopy radiation balance, i.e., changes in downward/upward shortwave radiation (SWDOWN/SWUP). The surface albedo is the parameter that has the most significant impact on most of the meteorological variables. Higher surface albedo reflects more shortwave radiation and reduces net shortwave radiation (NSW) entering the urban canopy, which leads to a decrease in both TSK and T2M. Previous studies also suggest that high surface albedo contributes significantly to reducing the TSK and canopy layer UHI [78,79]. For the effects of the three structural parameters, the effect of building height is opposite to those of roof width and road width, which is also consistent with the findings above (Figures 9–11). The increase in building height (roof width/road width) leads to more (less) net radiation (RN) in the urban canopy and a decrease (increase) in TSK due to the longer (shorter) building shadow. The temperature gradient between the canopy and ground increases and causes more energy to be transported into the ground (GRDFLX), resulting in a decrease in T2M. On the other side, the heat capacity and thermal conductivity have a greater influence on the surface energy balance, i.e., changes in HFX and GRDFLX. Their increases mean that the building could store more energy in the urban canopy and conduct it into the ground, therefore causing a negative difference in both HFX and GRDFLX. The increase in roughness length enhances the mechanical turbulence of the boundary layer, causing an increase in UST. More heat (RN) is conducted to the upper boundary layer during the daytime, resulting in an increase in GRDFLX and a decrease in TSK. The anthropogenic heat is treated as an additional sensible heat source in the model, and its impact is relatively small compared to the other UCPs in this study.

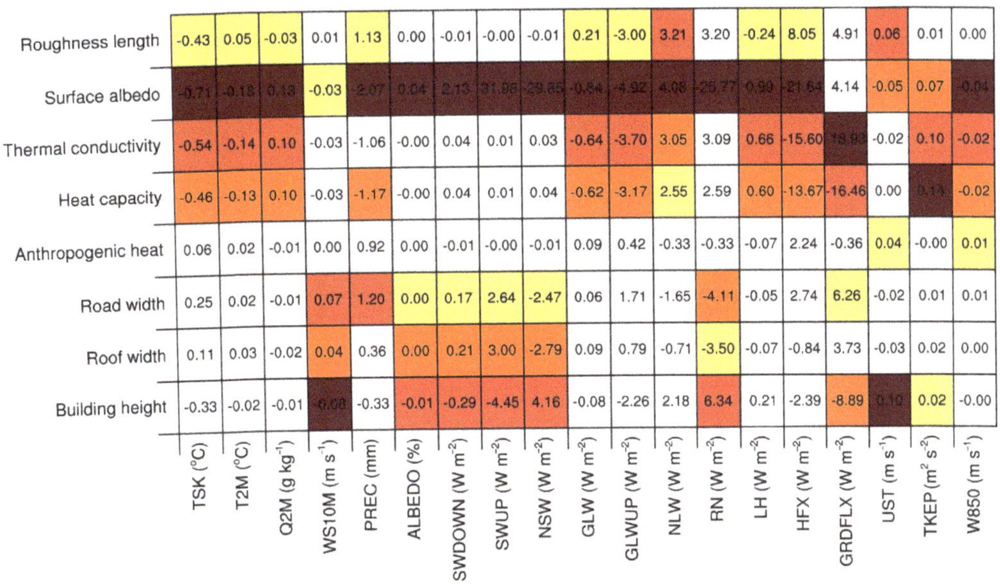

Figure 13. Differences in multiple meteorological variables averaged in the control region caused by increasing urban canopy parameters by 50% in each sensitivity test (i.e., SEN1, SEN2, ..., SEN16). The colored grid points are the four most changed members in each column. The darker the color, the greater the absolute value. TSK—surface temperature, T2M—2 m temperature, Q2M—2 m mixing ratio, WS10M—10 m wind speed, PREC—precipitation, ALBEDO—surface albedo, SWDOWN—downward shortwave radiation, SWUP—upward shortwave radiation, NSW—net shortwave radiation, GLW—downward longwave radiation, GLWUP—upward longwave radiation, NLW—net longwave radiation, RN—net radiation, LH—latent heat flux, HFX—sensible heat flux, GRDFLX—ground heat flux, UST—friction velocity, TKEP—turbulent kinetic energy flux at 850 hPa, W850—vertical velocity at 850 hPa.

4. Discussion

Local heavy precipitation is a common convective weather event in the summer on the east coast of China. However, since it usually occurs on a small scale and for a short period (i.e., less than 6 h), it is difficult to forecast and provide an early warning, and, thus, it poses a great challenge to daily city management. From the three sets of sensitivity experiments in this study, it can be seen that there are more uncertainties in forecasting precipitation in the urban area than in the other three meteorological variables (T2M, Q2M, and WS10M). This uncertainty comes from several sources. Firstly, the parameterization schemes introduce significant uncertainty in the simulation of precipitation. The uncertainties related to parameterization schemes arise from either uncertainty in the parameter or the parameterization [55]. A comparison of the parameterization schemes suggests that the microphysics scheme plays the most important role in forecasting precipitation, followed by the PBL, the land surface, and the urban physics schemes. This result is consistent with the work of Yu et al. [55] and Chen et al. [68], who also indicated that the microphysics and PBL schemes can substantially affect the temperature and moisture profiles in the lower troposphere and then have a significant impact on precipitation simulation. Secondly, the uncertainties brought about by the urban land use on precipitation are not limited to the urban area but extend to surrounding areas as well. In addition, unlike the other three meteorological variables, the effect of urbanization on precipitation shows a positive and negative staggered distribution. This may be because precipitation is the result of various meteorological elements (temperature, moisture, and wind convergence), and the impact of LCLU and UCPs on each meteorological element may sometimes be contradictory to changes in precipitation. Hence, precipitation changes may offset each other with different changes in UCPs, resulting in little changes in mean precipitation.

Although the model results are generally acceptable, it should be noted that the uncertainty in the initial field is not investigated in this study. The initial and boundary conditions were downscaled from global models (ERA5 reanalysis) with no other data assimilation or nudging performed. We believe more efforts could be made to better remove the uncertainties in the initial field, such as coupling HRLDAS to WRF [80], adjusting the initial time, and generating ensemble members by perturbing the initial field [81]. In addition, urbanization in the current study refers to a change in land use and UCPs but does not consider the impact of aerosols. The UCP values used herein are the default values in the WRF model. Future studies should consider more realistic high-resolution UCPs and atmospheric chemistry models to examine the effect of anthropogenic heat and aerosols on local heavy precipitation in cities.

5. Summary and Conclusions

A typical summer heavy rainfall event that occurred in Hangzhou, China, was chosen to analyze the sensitivity of UCPs and the individual and combined impacts of the thermal and dynamical effects of urbanization using the integrated WRF/urban modelling system.

The performance of the model was evaluated by using ensemble simulations, which included 24 members with different combinations of physical parameterization schemes (ENCTL). The results show that the ENCTL basically captures the heavy precipitation and the high temperature, low mixing ratio, and weak wind speed before precipitation, although many ensemble members underestimate the mixing ratio, overestimate the wind speed, and the convective initiation starts 0–2 h earlier than the observation.

The individual and combined impacts of the thermal and dynamical effects of urbanization were also examined. In this case, urbanization could lead to higher temperatures, lower mixing ratios, lower wind speed, and more precipitation in and around the urban area. Meanwhile, urban land use introduced more uncertainties than natural land use in forecasting the four meteorological variables, especially for precipitation. Both the thermal and dynamical effects contribute to an increase in temperature and precipitation, and the thermal effect plays a major role. The thermal and dynamical effects are the opposite of changes in the mixing ratio and wind speed, and they play a major role in each, respectively.

Eight important UCPs characterizing the urban canopy were chosen to investigate their impacts on this rainfall event before and when it began. The influences of UCPs are mainly concentrated in the urban area, although the effects of UCPs on precipitation are less evenly distributed in space than on the other three variables. The temperature, mixing ratio, and precipitation are more sensitive to changes in heat capacity, thermal conductivity, surface albedo, and roughness length, while wind speed is mainly affected by structural parameters (i.e., building height, roof width, and road width). In addition, the influences of UCPs on the urban canopy radiation balance and the surface energy balance were further explored.

Supplementary Materials: The following supporting information can be downloaded at: https://www.mdpi.com/article/10.3390/land12111965/s1, Figure S1: Differences in (a1–a3) surface skin temperature (°C), (b1–b3) surface net radiation (W m^{-1}), (c1–c3) ground heat flux (W m^{-2}), (d1–d3) sensible heat flux (W m^{-2}), and (e1–e3) latent heat flux (W m^{-1}) averaged during 10:00 to 13:00 BST on 26 July 2018 between the control test (CTL) and (left) NoUB, (middle) NoTH, and (right) NoDY test; Figure S2: Same as Figure S1 but for (a1–a3) friction velocity UST (m s^{-1}) and (b1–b3) turbulent kinetic energy TKEP (m^2 s^{-2}) at 850h Pa during 10:00 to 18:00 BST on 26 July 2018.

Author Contributions: Conceptualization, M.W. and F.C.; methodology, M.D.; software, F.C.; validation, M.W., F.C. and X.Y.; formal analysis, M.W.; investigation, M.W.; resources, F.C.; data curation, M.D.; writing—original draft preparation, M.W.; writing—review and editing, M.W. and F.C.; visualization, M.W. and F.C.; supervision, X.Y.; project administration, F.C.; funding acquisition, F.C. All authors have read and agreed to the published version of the manuscript.

Funding: This study was supported by the National Natural Science Foundation of China (grants 42030610, 42205006, and 41971019); Zhejiang Provincial Natural Science Foundation of China (LY21D050001); the Open Grants of the State Key Laboratory of Severe Weather (2021 LASW-B21).

Data Availability Statement: The ERA5 reanalysis dataset is available at https://cds.climate.copernicus.eu/cdsapp#!/home (accessed on 18 September 2023); the WRF-ARW coupled with urban model version 4.0.2 is downloaded from http://www2.mmm.ucar.edu/wrf/users (accessed on 1 May 2021). The temperature, mixing ratio, wind field, and temperature data were provided by the Zhejiang Meteorological Bureau, China Meteorological Administration, which are available from the authors upon reasonable request.

Conflicts of Interest: The authors declare no conflict of interest.

Abbreviations

Abbreviations and Descriptions

ALBEDO	surface albedo
GLW	downward longwave radiation
GLWUP	upward longwave radiation
GRDFLX	ground heat flux
HFX	sensible heat flux
LH	latent heat flux
NLW	net longwave radiation
NSW	net shortwave radiation
PBL	planetary boundary layer scheme
PREC	precipitation
Q2M	2 m mixing ratio
RMSE	root-mean-square error
RN	net radiation
SWDOWN	downward shortwave radiation
SWUP	upward shortwave radiation
TKEP850	turbulent kinetic energy flux at 850 hPa
TS	threat score
TSK	surface temperature
T2M	2 m temperature
UCP	urban canopy parameter
UHI	urban heat island
UST	friction velocity
WS10M	10 m wind speed
W850	vertical velocity at 850 hPa

References

1. Han, J.; Baik, J.; Lee, H. Urban impacts on precipitation. *Asia-Pac. J. Atmos. Sci.* **2014**, *50*, 17–30. [CrossRef]
2. Portman, D. Identifying and Correcting Urban Bias in Regional Time Series: Surface Temperature in China's Northern Plains. *J. Clim.* **1993**, *6*, 2298–2308. [CrossRef]

3. Ren, G.; Zhou, Y.; Chu, Z.; Zhou, J.; Zhang, A.; Guo, J.; Liu, X. Urbanization effects on observed surface air temperature trends in north China. *J. Clim.* **2008**, *21*, 1333–1348. [CrossRef]
4. Sun, X.; Luo, Y.; Gao, X.; Wu, M.; Li, M.; Huang, L.; Zhang, D.-L.; Xu, H. On the Localized Extreme Rainfall over the Great Bay Area in South China with Complex Topography and Strong UHI Effects. *Mon. Wea. Rev.* **2021**, *149*, 2777–2801. [CrossRef]
5. Jiang, X.; Luo, Y.; Zhang, D.; Wu, M. Urbanization Enhanced Summertime Extreme Hourly Precipitation over the Yangtze River Delta. *J. Clim.* **2020**, *33*, 5809–5826. [CrossRef]
6. Wen, J.; Chen, J.; Lin, W.; Jiang, B.; Xu, S.; Lan, J. Impacts of Anthropogenic Heat Flux and Urban Land-Use Change on Frontal Rainfall near Coastal Regions: A Case Study of a Rainstorm over the Pearl River Delta, South China. *J. Appl. Meteorol. Clim.* **2020**, *59*, 363–379. [CrossRef]
7. Wang, X.; Lin, W.; Yang, L.; Deng, R.; Lin, H. A numerical study of influences of urban land-use change on ozone distribution over the Pearl River Delta Region, China. *Tellus B Chem. Phys. Meteorol.* **2007**, *59*, 633–641. [CrossRef]
8. Zhang, H.; Wang, S.; Hao, J.; Wang, X.; Wang, S.; Chai, F.; Li, M. Air pollution and control action in Beijing. *J. Clean. Prod.* **2016**, *112*, 1519–1527. [CrossRef]
9. Mishra, V.; Dennis, P. Climatic trends in major U.S. urban areas, 1950–2009. *Geophys. Res. Lett.* **2011**, *38*. [CrossRef]
10. Westra, S.; Alexander, L.; Zwiers, F. Global increasing trends in annual maximum daily precipitation. *J. Clim.* **2013**, *26*, 3904–3918. [CrossRef]
11. Zhang, D. Rapid urbanization and more extreme rainfall events. *Sci. Bull.* **2020**, *65*, 516–518. [CrossRef] [PubMed]
12. Shepherd, J. A review of current investigations of urban-induced rainfall and recommendations for the future. *Earth Interact.* **2005**, *9*, 1–27. [CrossRef]
13. Liang, X.; Miao, S.; Li, J.; Bornstein, R.; Zhang, X.; Gao, Y.; Chen, F.; Cao, X.; Cheng, Z.; Clements, C.; et al. SURF: Understanding and predicting urban convection and haze. *Bull. Amer. Meteor. Soc.* **2018**, *99*, 1391–1413. [CrossRef]
14. Chen, F.; Yu, B.; Wu, M.; Yang, X. Improved Urban Finescale Forecasting During a Heat Wave by Using High-Resolution Urban Canopy Parameters. *Front. Clim.* **2022**, *3*, 771441. [CrossRef]
15. Miao, Y.; Liu, S.; Zheng, Y.; Wang, S.; Chen, B. Numerical Study of the Effects of Topography and Urbanization on the Local Atmospheric Circulations over the Beijing-Tianjin-Hebei, China. *Adv. Meteor.* **2015**, *2015*, 397070. [CrossRef]
16. Shui, T.; Liu, J.; Zhang, P.; Liu, S.; Zhao, Z. Development of an urban canopy model for the evaluation of urban thermal climate with snow cover in severe cold regions. *Build. Environ.* **2016**, *95*, 160–170. [CrossRef]
17. Kusaka, H.; Kimura, F. Coupling a single-layer urban canopy model with a simple atmospheric model: Impact on urban heat island simulation for an idealized case. *J. Meteor. Soc. Jpn.* **2004**, *82*, 67–80. [CrossRef]
18. Kusaka, H.; Kondo, H.; Kikegawa, Y.; Kimura, F. A simple single-layer urban canopy model for atmospheric models: Comparison with multi-layer and slab models. *Bound-Layer Meteor.* **2001**, *101*, 329–358. [CrossRef]
19. Martilli, A.; Clappier, A.; Rotach, M. An urban surface exchange parameterisation for mesoscale models. *Bound-Layer Meteor.* **2002**, *104*, 261–304. [CrossRef]
20. Chen, F.; Kusaka, H.; Bornstein, R.; Ching, J.; Grimmond, S.; Grossman-Clarke, S.; Loridan, T.; Manning, K.W.; Martilli, A.; Miao, S.; et al. The integrated WRF/urban modelling system: Development, evaluation, and applications to urban environmental problems. *Int. J. Climatol.* **2011**, *31*, 273–288. [CrossRef]
21. Liu, Y.; Chen, F.; Warner, T.; Basara, J. Verification of a mesoscale data-assimilation and forecasting system for the Oklahoma City area during the Joint Urban 2003 field project. *J. Appl. Meteorol. Clim.* **2006**, *45*, 912–929. [CrossRef]
22. Ganeshan, M.; Murtugudde, R.; Imhoff, M. A multi-city analysis of the UHI-influence on warm season rainfall. *Urban Clim.* **2013**, *6*, 1–23. [CrossRef]
23. Changnon, S. Rainfall changes in summer caused by St. Louis. *Science* **1979**, *205*, 402–404. [CrossRef] [PubMed]
24. Shepherd, J.; Harold, P.; Negri, A. Rainfall modification by major urban areas: Observations from spaceborne rain radar on the TRMM satellite. *J. Appl. Meteor.* **2002**, *41*, 689–701. [CrossRef]
25. Niyogi, D.; Lei, M.; Kishtawal, C.; Schmid, P.; Shepherd, M. Urbanization impacts on the summer heavy rainfall climatology over the eastern United States. *Earth Interact.* **2017**, *21*, 1–7. [CrossRef]
26. Bornstein, R.; Lin, Q. Urban heat islands and summertime convective thunderstorms in Atlanta: Three case studies. *Atmos. Environ.* **2000**, *34*, 507–516. [CrossRef]
27. Dou, J.; Wang, Y.; Bornstein, R.; Miao, S. Observed spatial characteristics of Beijing urban climate impacts on summer thunderstorms. *J. Appl. Meteorol. Clim.* **2015**, *54*, 94–105. [CrossRef]
28. Wan, H.; Zhong, Z.; Yang, X.; Li, X. Ensembles to model the impact of urbanization for a summertime rainstorm process in Yangtze River Delta, China. *Meteorol. Appl.* **2015**, *22*, 105–112. [CrossRef]
29. Miao, S.; Chen, F.; Li, Q.; Fan, S. Impacts of Urban Processes and Urbanization on Summer Precipitation: A Case Study of Heavy Rainfall in Beijing on 1 August 2006. *J. Appl. Meteorol. Clim.* **2011**, *50*, 806–825. [CrossRef]
30. Dou, J.; Bornstein, R.; Miao, S.; Sun, J.; Zhang, Y. Observation and simulation of a bifurcating thunderstorm over Beijing. *J. Appl. Meteorol. Clim.* **2020**, *59*, 2129–2148. [CrossRef]
31. Xian, T.; Guo, J.; Zhao, R.; Su, T.; Li, Z. The impact of urbanization on mesoscale convective systems in the Yangtze River Delta region of China: Insights gained from observations and modeling. *J. Geophys. Res. Atmos.* **2023**, *128*, e2022JD037709. [CrossRef]

32. Zhong, S.; Qian, Y.; Zhao, C.; Leung, R.; Wang, H.; Yang, B.; Fan, J.; Yan, H.; Yang, X.-Q.; Liu, D. Urbanization-induced urban heat island and aerosol effects on climate extremes in the Yangtze River Delta region of China. *Atmos. Chem. Phys.* **2017**, *17*, 5439–5457. [CrossRef]
33. Luo, Y.; Zhang, J.; Yu, M.; Liang, X.; Xia, R.; Gao, Y.; Gao, X.; Yin, J. On the Influences of Urbanization on the Extreme Rainfall over Zhengzhou on 20 July 2021, A Convection-Permitting Ensemble Modeling Study. *Adv. Atmos. Sci.* **2023**, *40*, 393–409. [CrossRef]
34. Chen, L.; Ng, E. Quantitative urban climate mapping based on a geographical database: A simulation approach using Hong Kong as a case study. *Int. J. Appl. Earth Obs. Geoinf.* **2011**, *13*, 586–594. [CrossRef]
35. Martilli, A. Current research and future challenges in urban mesoscale modelling. *Int. J. Climatol.* **2007**, *27*, 1909–1918. [CrossRef]
36. He, X.; Li, Y.; Wang, X.; Chen, L.; Yu, B.; Zhang, Y.; Miao, S. High-resolution dataset of urban canopy parameters for Beijing and its application to the integrated WRF/Urban modelling system. *J. Clean. Prod.* **2019**, *208*, 373–383. [CrossRef]
37. Ching, J.; Brown, M.; Burian, S. National Urban Database and Access Portal Tool. *Bull. Am. Meteor. Soc.* **2009**, *90*, 1157–1168. [CrossRef]
38. Baik, J.; Kim, Y.; Chun, H. Dry and moist convection forced by an urban heat island. *J. Appl. Meteor.* **2001**, *40*, 1462–1475. [CrossRef]
39. Kaufmann, R.; Seto, K.; Schneider, A.; Liu, Z.; Zhou, L.; Wang, W. Climate response to rapid urban growth: Evidence of a human-induced precipitation deficit. *J. Clim.* **2007**, *20*, 2299–2306. [CrossRef]
40. Coutts, A.; Daly, E.; Beringer, J.; Tapper, N. Assessing practical measures to reduce urban heat: Green and cool roofs. *Build. Environ.* **2013**, *70*, 266–276. [CrossRef]
41. Liu, X.; Tian, G.; Feng, J.; Wang, J.; Kong, L. Assessing summertime urban warming and the cooling efficacy of adaptation strategy in the Chengdu-Chongqing metropolitan region of China. *Sci. Total Environ.* **2018**, *610–611*, 1092–1102. [CrossRef] [PubMed]
42. Zhou, X.; Chen, H. Impact of urbanization-related land use land cover changes and urban morphology changes on the urban heat island phenomenon. *Sci. Total Environ.* **2018**, *635*, 1467–1476. [CrossRef] [PubMed]
43. Zonato, A.; Martilli, A.; Di Sabatino, S.; Zardi, D.; Giovannini, L. Evaluating the performance of a novel WUDAPT averaging technique to define urban morphology with mesoscale models. *Urban Clim.* **2020**, *31*, 100584. [CrossRef]
44. Yu, M.; Chen, X.; Yang, J.; Miao, S. A new perspective on evaluating high-resolution urban climate simulation with urban canopy parameters. *Urban Clim.* **2021**, *38*, 100919. [CrossRef]
45. Shen, C.; Chen, X.; Dai, W.; Li, X.; Wu, J.; Fan, Q.; Wang, X.; Zhu, L.; Chan, P.; Hang, J.; et al. Impacts of High-Resolution Urban Canopy Parameters within the WRF Model on Dynamical and Thermal Fields over Guangzhou, China. *J. Appl. Meteorol. Clim.* **2019**, *58*, 1155–1176. [CrossRef]
46. Zhang, C.; Chen, F.; Miao, S.; Li, Q.; Xia, X.; Xuan, C. Impacts of urban expansion and future green planting on summer precipitation in the Beijing metropolitan area. *J. Geophys. Res. Atmos.* **2009**, *114*, D02116. [CrossRef]
47. Zhong, S.; Qian, Y.; Zhao, C.; Leung, R.; Yang, X. A case study of urbanization impact on summer precipitation in the Greater Beijing Metropolitan Area: Urban heat island versus aerosol effects. *J. Geophys. Res. Atmos.* **2015**, *120*, 10903–10914. [CrossRef]
48. Wang, X.; Sun, X.; Tang, J.; Yang, X. Urbanization-induced regional warming in yangtze river delta: Potential role of anthropogenic heat release. *Int. J. Climatol.* **2015**, *35*, 4417–4430. [CrossRef]
49. Jiang, X.; Zhang, D.; Luo, Y. Influences of Urbanization on an Afternoon Heavy Rainfall Event over the Yangtze River Delta Region. *Mon. Wea. Rev.* **2023**, *151*, 815–832. [CrossRef]
50. Wu, M.; Luo, Y.; Chen, F.; Wong, W. Observed Link of Extreme Hourly Precipitation Changes to Urbanization over Coastal South China. *J. Appl. Meteor. Clima.* **2019**, *58*, 1799–1819. [CrossRef]
51. Li, Y.; Wang, W.; Chang, M.; Wang, X. Impacts of urbanization on extreme precipitation in the Guangdong-Hong Kong-Macau Greater Bay Area. *Urban Clim.* **2021**, *38*, 100904. [CrossRef]
52. Chen, F.; Yang, X.; Wu, J. Simulation of the urban climate in a Chinese megacity with spatially heterogeneous anthropogenic heat data. *J. Geophys. Res. Atmos.* **2016**, *121*, 5193–5212. [CrossRef]
53. Skamarock, W.; Klemp, J.; Dudhia, J.; Gill, D.O.; Liu, Z.; Berner, J.; Huang, X.Y. *A Description of the Advanced Research WRF Version 4*; NCAR Technical Notes NCAR/TN-556+STR 2019; UCAR: Boulder, CO, USA, 2019; pp. 145p.
54. Berner, J.; Ha, S.; Hacker, J.; Fournier, A.; Snyder, C. Model uncertainty in a mesoscale ensemble prediction system: Stochastic versus multiphysics representations. *Mon. Wea. Rev.* **2011**, *139*, 1972–1995. [CrossRef]
55. Yu, M.; Miao, S.; Zhang, H. Uncertainties in the impact of urbanization on heavy rainfall: Case study of a rainfall event in Beijing on 7 August 2015. *J. Geophys. Res. Atmos.* **2018**, *123*, 6005–6021. [CrossRef]
56. Iacono, M.; Delamere, J.; Mlawer, E.; Shephard, M.; Clough, S.; Collins, W. Radiative forcing by long-lived greenhouse gases: Calculations with the AER radiative transfer models. *J. Geophys. Res. Atmos.* **2008**, *113*, D13103. [CrossRef]
57. Chen, S.; Sun, W. A One-dimensional Time Dependent Cloud Model. *J. Meteor. Soc. Jpn. Ser. II* **2002**, *80*, 99–118. [CrossRef]
58. Hong, S.; Lim, J. The WRF single-moment 6-class microphysics scheme (WSM6). *J. Korean Meteor. Soc.* **2006**, *42*, 129–151.
59. Lim, K.; Hong, S. Development of an Effective Double-Moment Cloud Microphysics Scheme with Prognostic Cloud Condensation Nuclei (CCN) for Weather and Climate Models. *Mon. Wea. Rev.* **2010**, *138*, 1587–1612. [CrossRef]
60. Janjić, Z. The Step-Mountain Eta Coordinate Model: Further Developments of the Convection, Viscous Sublayer, and Turbulence Closure Schemes. *Mon. Wea. Rev.* **1994**, *122*, 927–945. [CrossRef]
61. Mesinger, F. Forecasting upper tropospheric turbulence within the framework of the Mellor Yamada 2.5 closure. *Res. Activ. Atmos. Oceanic Mod.* **1993**. Available online: https://www.researchgate.net/profile/Fedor-Mesinger/publication/34361084

9_Forecasting_upper_tropospheric_turbulence_within_the_framework_of_the_Mellor-Yamada_25_closure/links/5f341119458515b7291bc95c/Forecasting-upper-tropospheric-turbulence-within-the-framework-of-the-Mellor-Yamada-25-closure.pdf (accessed on 1 July 2018).
62. Bougeault, P.; Lacarrere, P. Parameterization of Orography-Induced Turbulence in a Mesobeta-Scale Model. *Mon. Wea. Rev.* **1989**, *117*, 1872–1890. [CrossRef]
63. Tewari, M.; Chen, F.; Wang, W.; Dudhia, J.; LeMone, M.A.; Mitchell, K.; Cuenca, R.H. Implementation and verification of the united NOAH land surface model in the WRF model. In Proceedings of the 20th Conference on Weather Analysis and Forecasting/16th Conference on Numerical Weather Prediction, Seattle, WA, USA, 12–16 January 2004; Volume 1115, pp. 2165–2170.
64. Niu, G.; Yang, Z.; Mitchell, K.; Chen, F.; Ek, M.B.; Barlage, M.; Xia, Y. The community Noah land surface model with multiparameterization options (Noah-MP): 1. Model description and evaluation with local-scale measurements. *J. Geophys. Res. Atmos.* **2011**, *116*, D12109. [CrossRef]
65. Hersbach, H.; Bell, B.; Berrisford, P.; Dahlgren, P.; Horányi, A.; Munoz-Sebater, J.; Soci, C. The ERA5 Global Reanalysis: Achieving a detailed record of the climate and weather for the past 70 years. *Eur. Geophys. Union Gen. Assem.* **2020**, 3–8. Available online: https://ui.adsabs.harvard.edu/abs/2020EGUGA..2210375H/abstract (accessed on 12 July 2021).
66. Strahler, A. MODIS Land Cover Product Algorithm Theoretical Basis Document (ATBD) Version 5.0. *Tech. Rep.* **1999**, *72*, 42–47.
67. Chen, F.; Yang, X.; Zhu, W. WRF simulations of urban heat island under hot-weather synoptic conditions: The case study of Hangzhou City, China. *Atmos. Res.* **2014**, *138*, 364–377. [CrossRef]
68. Chen, F.; Wu, M.; Dong, M.; Yu, B. Comparison of the Impacts of Topography and Urbanization on an Extreme Rainfall Event in the Hangzhou Bay Region. *J. Geophys. Res. Atmos.* **2022**, *127*, e2022JD037060. [CrossRef]
69. Ikeda, K.; Rasmussen, R.; Liu, C.; Gochis, D.; Yates, D.; Chen, F.; Tewari, M.; Barlage, M.; Dudhia, J.; Miller, K.; et al. Simulation of seasonal snowfall over Colorado. *Atmos. Res.* **2010**, *97*, 462–477. [CrossRef]
70. Xu, L.; Dong, M.; Chen, F. Comparison study of spatial interpolation methods based on hourly precipitation data from automatic weather stations. *J. Meteor. Environ.* **2017**, *33*, 34–43. (In Chinese)
71. Gaur, A.; Lacasse, M.; Armstrong, M.; Lu, H.; Shu, C.; Fields, A.; Palou, F.S.; Zhang, Y. Effects of using different urban parametrization schemes and land-cover datasets on the accuracy of WRF model over the City of Ottawa. *Urban Clim.* **2021**, *35*, 100737. [CrossRef]
72. Guo, X.; Fu, D.; Wang, J. Mesoscale convective precipitation system modified by urbanization in Beijing City. *Atmos. Res.* **2006**, *82*, 112–126. [CrossRef]
73. Xing, Y.; Ni, G.; Yang, L.; Yang, Y.; Xing, P.; Sun, T. Modeling the impacts of urbanization and open water surface on heavy convective rainfall: A case study over the emerging Xiong'an City, China. *J. Geophys. Res. Atmos.* **2019**, *124*, 9078–9098. [CrossRef]
74. Carter, M.; Shepherd, J.; Burian, S.; Jeyachandran, I. Integration of lidar data into a coupled mesoscale–land surface model: A theoretical assessment of sensitivity of urban–coastal mesoscale circulations to urban canopy parameters. *J. Atmos. Ocean. Technol.* **2012**, *29*, 328–346. [CrossRef]
75. Buizza, R. Potential Forecast Skill of Ensemble Prediction and Spread and Skill Distributions of the ECMWF Ensemble Prediction System. *Mon. Wea. Rev.* **1997**, *125*, 99–119. [CrossRef]
76. Miao, S.; Chen, F.; LeMone, M.; Tewari, M.; Li, Q.; Wang, Y. An Observational and Modeling Study of Characteristics of Urban Heat Island and Boundary Layer Structures in Beijing. *J. Appl. Meteor. Clima.* **2009**, *48*, 484–501. [CrossRef]
77. Sun, Y.; Zhang, N.; Miao, S.; Kong, F.; Zhang, Y.; Li, N. Urban Morphological Parameters of the Main Cities in China and Their Application in the WRF Model. *J. Adv. Model. Earth Syst.* **2021**, *13*, e2020MS002382. [CrossRef]
78. Salamanca, F.; Martilli, A.; Yagüe, C. A numerical study of the urban heat island over madrid during the DESIREX (2008) campaign with WRF and an evaluation of simple mitigation strategies. *Int. J. Climatol.* **2012**, *32*, 2372–2386. [CrossRef]
79. Mughal, M.; Li, X.; Norford, L. Urban heat island mitigation in Singapore: Evaluation using WRF/multilayer urban canopy model and local climate zones. *Urban Clim.* **2020**, *34*, 100714. [CrossRef]
80. Chen, F.; Manning, K.; LeMone, M.; Trier, S.B.; Alfieri, J.G.; Roberts, R.; Tewari, M.; Niyogi, D.; Horst, T.W.; Oncley, S.P.; et al. Description and evaluation of the characteristics of the NCAR high-resolution land data assimilation system. *J. Appl. Meteor. Climatol.* **2007**, *46*, 694–713. [CrossRef]
81. Chen, J.; Xue, J.; Yan, H. A Comparison of a New Initial Condition and Model Perturbation Method for Heavy Rainfall Ensemble Prediction. *Acta Meteor. Sin.* **2009**, *23*, 53–67.

Disclaimer/Publisher's Note: The statements, opinions and data contained in all publications are solely those of the individual author(s) and contributor(s) and not of MDPI and/or the editor(s). MDPI and/or the editor(s) disclaim responsibility for any injury to people or property resulting from any ideas, methods, instructions or products referred to in the content.

MDPI AG
Grosspeteranlage 5
4052 Basel
Switzerland
Tel.: +41 61 683 77 34

Land Editorial Office
E-mail: land@mdpi.com
www.mdpi.com/journal/land

Disclaimer/Publisher's Note: The title and front matter of this reprint are at the discretion of the Guest Editors. The publisher is not responsible for their content or any associated concerns. The statements, opinions and data contained in all individual articles are solely those of the individual Editors and contributors and not of MDPI. MDPI disclaims responsibility for any injury to people or property resulting from any ideas, methods, instructions or products referred to in the content.

www.ingramcontent.com/pod-product-compliance
Lightning Source LLC
Chambersburg PA
CBHW062053220926
51850CB00010B/2753